Palgrave Advances in the Economics of Innovation and Technology

Series Editor
Albert N. Link
University of North Carolina at Greensboro
Greensboro, NC, USA

The focus of this series is on scholarly inquiry into the economic foundations of technologies and the market and social consequences of subsequent innovations. While much has been written about technology and innovation policy, as well as about the macroeconomic impacts of technology on economic growth and development, there remains a gap in our understanding of the processes through which R&D funding leads to successful (and unsuccessful) technologies, how technologies enter the market place, and factors associated with the market success (or lack of success) of new technologies.

This series considers original research into these issues. The scope of such research includes in-depth case studies; cross-sectional and longitudinal empirical investigations using project, firm, industry, public agency, and national data; comparative studies across related technologies; diffusion studies of successful and unsuccessful innovations; and evaluation studies of the economic returns associated with public investments in the development of new technologies.

More information about this series at
http://www.palgrave.com/gp/series/14716

Caroline S. Wagner

The Collaborative Era in Science

Governing the Network

palgrave
macmillan

Caroline S. Wagner
Glenn College of Public Affairs
The Ohio State University
Columbus, OH, USA

Palgrave Advances in the Economics of Innovation and Technology
ISBN 978-3-030-06948-3   ISBN 978-3-319-94986-4   (eBook)
https://doi.org/10.1007/978-3-319-94986-4

Cover image © A-Digit/Getty Images
Cover design by Alice Tomlinson and Laura de Grasse

This Palgrave Macmillan imprint is published by the registered company Springer Nature Switzerland AG
The registered company address is: Gewerbestrasse 11, 6330 Cham, Switzerland

PREFACE: TWENTY-FIRST-CENTURY SCIENCE

This book describes the global network of scientific collaboration that has emerged in the 21st century. The network is one part of the innovation process and feeds other parts of the larger knowledge economy. The book is written as an aid to those who contribute to the network through science policy, as they seek to use it or to evaluate it. It tells the story of the rapid rise of international collaborations in science, technology, and innovation which has been one of the most dramatic social changes of the twenty-first century. The growth to scientific maturity of many developing countries, especially China, has reshaped the global knowledge system and created the conditions for the global network to emerge. The extra-national nature of collaboration creates a new dynamic for policymakers, educators, and business people who fund, manage, and rely upon this system for a long list of social, economic, and knowledge benefits.

The growth of global science has exceeded the expectations of sociologists and science policy analysts and now constitutes a social system in its own right. Mid-century students of science, thrilled with the rapid progress of science, could not imagine a future of exponential growth: Derek de Solla Price (1963) forecasted that growth in output of scientific papers would flatten out long before 2018. After all, in the physical world, no system grows exponentially unabated. Science has been growing exponentially for more than three centuries. Price predicted a point of "saturation" of scientific output, one that has never been reached. The production of scientific output in articles, patents, and products has blown through expected limits to growth. A graph of scientific output shows the "hockey stick" of exponential growth (made famous by Michael Mann et al. (1999) in describing climate change) as applying to scientific articles.

Science growth operates within a law of accelerating returns, described by Moravec (1998), which suggests that, when growth reaches a barrier, a new technology is invented to allow us to cross the barrier. Such may be said of recent developments in the conduct of science, in that the Internet and associated digital technologies overcame the barrier of article production that was previously limited by paper. The invention of digitized information, large-scale storage, and then the Internet has created the conditions by which scientific knowledge output has continued to grow at an accelerating pace. Digitization has also contributed to wider diffusion of knowledge as well, which has contributed to a virtuous cycle of capacity growth, especially in developing and underdeveloped nations.

Science has expanded beyond boundaries of institutions, nations, and the disciplines of the twentieth century. Along with this shift across barriers (from paper to digital) has gone an accompanying phase shift from scarcity to abundance. Until quite recently, scientific knowledge was a scarce and precious resource. In the recent past (as it has been for three centuries in the West), science required access to expensive journals. These were printed on paper, mailed to subscribers, and often bound in leather (being treated as the precious resource it was) for cataloging in elite libraries. Accessing this codified knowledge meant expensive and time-consuming travel, time away from home, relocation or collocation in universities or research laboratory, and long time periods of mailing materials across long distances. Much knowledge is now electronically available. Of the materials that remain behind subscription-based "pay walls" established to share journal articles, much of it is bulging at the seams of those walls and beginning to spill out through sharing sites such as ResearchGate, SciHub, and others.

ABUNDANCE REORGANIZES SCIENCE

The system created by the broad sharing of scientific knowledge operates as abundant system. As much as half of 2017 scientific articles were published in "open access" formats. Online, Google Scholar's search engine offers work-arounds to find articles that would otherwise be difficult to access. Many websites provide access to scientific data. Other web-based services such as arXiv have scientific preprints online. All these sources and services are new ways of sharing scientific knowledge.

The abundance of scientific knowledge available in digital form brings with it many new norms, rules, and expectations for both science and the mechanisms that support it. Policymakers and research managers are facing an unknown or at least untested landscape. The most notable change for them has been one pushing science toward openness, teaming, cooperation, and collaboration. In fact, this book argues that openness is the child of abundance, and it is the natural and expected development emerging from the many new sources of knowledge. Many other changes go along with these shifts. These are described in this book.

> *An old tradition and a new technology have converged to make possible an unprecedented public good. The old tradition is the willingness of scientists and scholars to publish the fruits of their research in scholarly journals without payment, for the sake of inquiry and knowledge. The new technology is the internet. The public good they make possible is the world-wide electronic distribution of the peer-reviewed journal literature and completely free and unrestricted access to it by all scientists, scholars, teachers, students, and other curious minds.* (BOAI 2001)

The movement led to the creation of the Creative Commons and the establishment of the Open Society Institute.

> *By "open access" ... we mean its free availability on the public internet, permitting any users to read, download, copy, distribute, print, search, or link to the full texts of these articles, crawl them for indexing, pass them as data to software, or use them for any other lawful purpose, without financial, legal, or technical barriers other than those inseparable from gaining access to the internet itself. The only constraint on reproduction and distribution, and the only role for copyright in this domain, should be to give authors control over the integrity of their work and the right to be properly acknowledged and cited.* (BOAI 2001)

Scientific collaborations grew rapidly as open sharing—a broader concept than "open access"—has grown into a norm of research behavior. Open sharing was facilitated by lower-cost and rapid travel, information technologies, and open data. Each of these features converged to enable distributed teaming by members who are geographically dispersed but intellectually connected. These dispersed communities are networks—they are a social gathering of connections that extend beyond one's home institution or discipline, engage others, exchange ideas, and share tasks. This process of connecting to colleagues, and then being reconnected by them to yet other colleagues, is known as "search," and it is increasingly seen as a fundamental part of knowledge creation. Global knowledge networks are the connections among researchers and research institutions that are becoming the dominant form of organization for research.

A global knowledge network has come to dominate science in the early twenty-first century, as I discussed in my book: *The New Invisible College: Science for Development* (2008). The global network is the product of the decisions of thousands of scientists to reach beyond the borders of laboratories within their home countries to seek out collaborators who can enhance the knowledge-creating process. Collaborations overcome the limitations imposed by organizational specialties. Long-distance collaborations overcome the limitations imposed by social encumbrances (like free riders and laggards). Collaborations bring new insights and possibilities that were not available at close proximity. The result has been to create a robust global knowledge network that is populated by prominent and productive scientists. This book moves beyond my earlier work to discuss the system at three levels (individual, team, and nation) and to present governing guidelines for them.

New knowledge is increasingly created within these international networks—the work tends to be more highly cited by other scientists, and funding is more often going to these projects rather than to solely national ones. As knowledge is created by dispersed groups, the questions of exactly where the knowledge "resides," who "owns" it, who "controls" it become issues for those who need to account for or use it. New entrants (businesses, students, those from developing countries) seeking to tap into this knowledge may be having a harder time, since it is widely dispersed. Finally, the global knowledge system operates like a network—with dynamics different from the hierarchies or bureaucracies that might have once been the homes of scientific research teams. It must be governed, and governance must ensure the welfare of those who ultimately pay for the public good to be created. To bring the knowledge from the global to the local users, and to have it work effectively, users must understand the emerging system and what motivates its participants.

Indeed, we find that political ties or national prestige do not motivate the alliances of researchers within the global knowledge network. Thus, the results of globally connected teams cannot be easily claimed as an asset by a single nation. Who "owns" the results of the Human Genome Project (HGP)? This six-nation project had hundreds of adherents who have gone on to use the resulting knowledge in myriad ways. In contrast to the secretive Manhattan Project of the 1940s, the HGP contributed knowledge as a "global public good" which seeded the genetics revolution. As a knowledge-creating system, the global knowledge network's breadth and scope is unprecedented in history. It is exceedingly influential in that elite scientists are populating it. As we shall see, the network operates according to its own internal dynamics—ones that are beyond the direct control of science policy makers but whose understanding will improve outcomes. This book describes this network and details the importance of the network for the future of scientific and technological research and development, and the implications of it for public policy.

Human organizations, like living organisms, evolve in response to environmental changes and challenges. In science, these challenges come in response to new information, data, and opportunities to confirm or develop new knowledge. New configurations and interconnections must be grown organically to adjust and adapt to new information—just as biochemistry emerged from the crossovers between biology and chemistry. Nanotechnology emerged from the capacities provided by electron tunneling microscopes; new opportunities propelled atomic physics, materials sciences, and chemistry in new directions and convergences. The new knowledge forms are potent and productive. They create entirely new opportunities for growth.

In response to these new signals, new organizational forms are emerging, often superseding (but not necessarily eradicating) the ones preceding them. In the current era, we have seen many global organizational forms emerge in finance, media, and business—but these do not remove the local or regional services. The global scientific network is one more global system—it has similar features to these other global enterprises, in that a new type of organization has emerged. However, it is also different from these in its origins and structures because it produces a public good—accordingly, its governance rules will differ from other global systems.

Like other human endeavors, scientific organization responds to external and internal changes working according to systemic mechanisms. Changes in one aspect of the system changes dynamics for all other parts. For twenty-first-century science, nearly all the parts have changed. One change has been the opportunity afforded by simply more scientists working around the world. The global network of science has emerged because scientists connect with each other on a peer-to-peer level, and a process of preferential attachment selects specific individuals into an increasingly elite circle. The process reduces free riders and greatly increases the visibility of parts of the system. The global network of science enhances the possibility that said research will be applied and cited or referenced by others. Joining the global network of science affords participants key benefits, as well as opportunities to passively exclude others who do not offer substantial contributions. It is the cream rising naturally to the top of the bottle. Policy is needed to reverse the natural tendency to have elites dominate the network.

In its reach and potential, the global network of science can be ranked among the most significant of human cooperative ventures. It is a coalition of highly trained people linked together to research the natural world; solve problems; share data, tools, and knowledge; all at a scale unheard of in human history. Inquiry into the natural world is being pursued farther from political agendas than at any time in the past, although political systems must learn to govern it properly. Ironically, no one would have predicted its rise, nor did any one set out to create it. In fact, by extending beyond the reach of governments, it becomes increasingly challenging to govern, and therefore perhaps less desirable to states, putting funding in jeopardy. The fact that the network appears to be influencing the governmental agenda rather than the other way around can only increase the political ambivalence about it. The challenge to governments is to learn to operate with and benefit from a network.

This book continues a thread of work treating science as a network. The suggestion that science operates as a network has been discussed by sociologists Robert Merton (1957) and Diana Crane (1972), philosophers of science Derek Price (1963) and Mary Hesse (1963), W.O. Hagstrom (1965) and bibliometrician Eugene Garfield (1972). The insights of these earlier observers were speculative: the network structure could not be tested at the scale and scope that is available now. Tools to visualize networks, and theories about their structure, have been greatly enhanced. Indeed, as we shall see in this book, one of the most amazing findings is that science communications networks share reproducible and universal characteristics—perhaps deep patterns of similarities—with many other types of networks. These similarities allow us to apply network theories and tools to science, and to use the resulting insights to inform policy.

In an effort to describe the emerging system, some have called it "Second Order Science" (Müller and Riegler 2014), Science 2.0 (Nentwich and König 2012), and the Fourth Age of Science (Adams 2013)—each attempting to describe parts of the emerging system. This book takes a fuller approach by detailing its formative dynamics, its structure, and the implications of these phenomena on public access and use of these critical aspects of knowledge in the realm of public policy. Policymakers are expected to maximize the benefits of the system—governing it requires understanding it.

John Padgett and Walter Powell (2012) argued that organizational innovation must reproduce itself in order to survive. To survive and thrive, any new organization must attract resources and train people. This is a striking feature of the global network: the most elite, highly productive scientists are populating it. It is highly attractive. In fact, the more elite the scientist, the more likely he or she is working collaboratively at the global level. The public funding agencies have shifted to favor international collaboration projects out of necessity, since this is where their top scientists are choosing to work. The global collaborations are increasingly team-based (working together toward a common goal), and very often involve the convergence of several disciplines, to approach a complex problem. Many factors within the knowledge system are influencing these changes; while we will address these factors, the fundamental prerequisite of this perspective is gaining understanding of the global network itself.

With the vitality and resources of a new organization, the global network of science has tremendous promise to solve critical problems facing humankind. In many ways, it is already doing so, as this book will detail, but a network does not solve problems or make decisions (in fact, networks are poor at decision-making). Public policy solves problems. As Charlotte Hess and Elinor Ostrom (2005) point out, "the more people who share useful knowledge, the greater the common good." (p. 5) But governing global science as a conglomeration of national assets or as disciplines, or attempting to renationalize science, threatens to choke off the promise embedded in the emerging system. The methods of influencing and using networks to solve human problems require significant changes in the policy processes, and these are eminently worth learning. Lessons learned from governing network structures will be adapted to the global network, with recommendations offered to maximize the benefits of network dynamics. Governance of the network is critical to ensuring fairness and inclusiveness. Thus the book explores those laws of action that, as Nancy Cartwright put it, "are fundamental and explanatory," to be distinguished from "those that merely describe."

MULTIPLE LEVELS OF ANALYSIS

While some claim that scientific ideas spark first in the mind of an individual, this book considers the group as much or more than the individual. Certainly, individuals operate within the network, but it is the social group (disciplines and practitioners) that validates knowledge. The validated knowledge is not private or subjective: it belongs to the group. Thus, the shared understanding is created socially in the synthesis of differing and additive views by a variety of people. The individuals express value judgments of new ideas based upon shared history and group norms. The collective is responsible for the creative process. Institutions retain knowledge created within that process. A multi-level analysis is needed, leading up to explaining the global level.

In addition to being a fascinating shift in human organization, the global network also represents a new stratum in the history of science. Paul David (2004) and other historians of science describe modern science—the activity emerging in the enlightenment and eventually tied to nation-states—as beginning in the seventeenth century. Individuals, such as Galileo or Bacon, are the early heroes, characterized as lone geniuses who operated outside of established institutions to seek enlightenment about fundamental questions in nature.[1] Systemic features (where science becomes a method of thinking, widely supported by society) emerge in the eighteenth century as laboratory-based research develops. As science proved useful to industry and government, its practice became a profession for which universities trained young people, and for which governments began providing support during the nineteenth century. In the twentieth century, science became a tool of nationalism, and its practice became increasingly tied to national prestige. New technology grew in the service of military power, and a means for economic growth and national competitiveness, further supporting investments in science.

---

[1] It is arguable that "lone geniuses" never really operated alone, but that the attribution of advances became associated with a single name. Even Isaac Newton claimed to have stood on the shoulders of giants. Collaboration may have a long history in science, but only recently has it become the practice to list multiple authors on publications.

The vestiges of each of these previous eras remain as layers within the current system (with the possible exception of the "lone heroes of science"), but the practice of science in the twenty-first century is undergoing a shift from the individual to the institutional, to the national, and now to the global network. The organization is a combination of historical features such as the invisible colleges described by Diana Crane (1972) or epistemic communities detailed by Karin Knorr-Cetina (2009), but with the added detail that the network renders less potent the hierarchy and protective boundaries of previous forms of organization. The new organizational form challenges nations to shift policies to retain and grow their scientific wealth (often embodied in talented people) through policy changes designed to take advantage of networked knowledge rather than institutionally based, politically protected centers.

The book focuses on those knowledge-creating processes that are conducted to be shared with others. The outputs represent knowledge processes of exploration, retention (through codification), and exploitation (Lichtenthaler and Lichtenthaler 2009) of useful knowledge, conducted for the growth of public good. Such knowledge may feed development in both the public and private sectors, but it is not "innovation" in the sense used by economists to describe products and processes new to the market, but in the sense of addressing a problem with new ideas, methods, or materials contributing to a global knowledge commons, in the sense suggested by Hess and Ostrom in *The Global Knowledge Commons* (2005).

Two aspects of the global network of science present the greatest challenges to governance, planning, and investment and thus are the focus of this book. One is that much more of the best research takes place at the international level, a fact shown through citation analysis and network structures, meaning new methods are needed to tap it. Scientific collaboration, joint investment, the mobility of researchers, and the global nature of problems are all influencing the shift—these features will be discussed in more detail. The second challenge is that the global network hosts more team-oriented and more interdisciplinary science, such that we now see a substantial growth in the collaborative practices across many fields, and the need for new methods to evaluate it. Even though these two phenomena are obviously increasing, they have not been well studied, and we lack a theory of practice to guide governance. These gaps are also addressed.

The emergence of the global system, grounded as it is in social networks—shifts science away from a national identity—or as part of what some may call the "national innovation system"—and offers the potential to improve efficiency in the research process and diffuse knowledge to many more users. There is a danger in it, too. The global order emerges from multiple, intertwined links; these, in turn, respond to a deep transformational dynamic based on both path-dependency and novelty (Barzel and Barabási 2013). Concerns about social justice, inclusiveness, and fairness are not a natural part of emergent network formation: *they must be added by public policy*. New, effective policies and management tools are needed to ensure inclusiveness of the global knowledge network benefits to all, particularly because we face extreme challenges to human welfare and environmental stability to which science can directly contribute. This book focuses with promise on the global network—a system which provides the opportunity to shift away from elitist, nationalist, and disciplinary bases that have arguably encumbered the broader usefulness of science over the past decades.

The self-organization of global science also presents the opportunity to study human cooperation. Why and how humans cooperate has been discussed for centuries in many disciplines. The global network self-organizes around the furtherance of the interest of its members, but the mechanism of reaching goals is collective social action. The global network is far from cutthroat competition or "coerced social good" postulated by some social thinkers. Far from being a competitive, dog-eat-dog world, the abundance of information, and the opportunity to share knowledge in science, has created an unprecedented level of human cooperation. This level of cooperation cannot be explained by history nor by theories of globalization or trans-nationality. It can, however, be explained by complex adaptive systems theory, which suggests that complex order can emerge from the self-interested actions of individuals both cooperating around and competing for resources on a landscape. The whole can emerge as greater than the sum of the parts. This book explores the governance of the global network of science with hope that others take up the implications for our understanding of human nature that may arise from examining this level of social organization.

The Meaning of Change

The changes discussed in this book arise from the spectacular growth of science and technology in networked form and its success in disseminating a way of conducting verifiable research and development characterizing the natural world. It is now widely accepted that science and technology underlie much of economic growth and social change. The social goods provided by scientific inquiry have been a highlight of human social development in the West for over 300 years. Science has grown along with social welfare and economy in most countries in a positive feedback loop. The most tangible parts of scientific growth have been an exponential rise in the stock of codified scientific knowledge (most tangibly obvious in articles, journals, newsletters, patents, and books) over the past three centuries, assisted by development of highly sophisticated and often large-scale scientific equipment (of which the Large Hadron Collider is only one example). It is hard not to notice that the growth of scientific output has occurred in line with the growth of the size and increasing success of the human race as a species—and also, with the impact of that species on the globe. Indeed, the vast numbers of people and the extended life span we enjoy are direct results of scientific advances. Helping all people live a decent life is our challenge now.

Thus, the white-coated scientist may labor on inside his or her laboratory (whether it is in a university, a government organization, or a private company), but the boundaries and structure of the system within which he or she works is changing rapidly. As we will see, the rules and incentives of the past are either defunct or no longer produce the same expected results. The changing system of science is altering the requirements for support, including such fundamental features as funding, patenting, allocating laboratory space, publishing, and knowledge sharing. The shift introduces new constraints on the one hand, and new incentives on the other, all built to encourage or control diffusion. The resulting mix is chaotic. Within the mix are invention disclosures and rights to a monopoly, imposition of export controls, and use of trade secrets. All of these features of the system are put to the test, and many policies are failing to show they can continue going forward to offer the benefits they once did. We will explore this in detail in the following chapters.

The thoughts we generate today and flick from mind to mind, as Lewis Thomas describes in The Lives of a Cell (1978). Thoughts transfer across space and time as communication from individuals that increasingly interweave into collectives. The collective is part of a co-evolving knowledge environment. Perhaps the interliving coral shoals of thought that Thomas prophesized are the multiple network typologies, which share the dynamics of the global system. The flickers of thought and the interliving shoals can be studied to enhance understanding. The nurturing of this more collective intelligence—the process toward E.O. Wilson's (1999) "consilience of knowledge"—is to enhance the journey toward an abundant knowledge system, one that offers greater promise to humankind in producing benefits from science in practicable ways.

PLAN OF THE BOOK

This book diverges from several trends in science policy, first by taking a social and administrative approach to science, rather than a "scientific" one. The last decade has seen a call for a "science of science policy"—a science which can never be created. Moreover, the practice of science takes place within the social sphere, and thus cannot be subject to scientific experimentation of the natural sciences. This book follows more the ideas of Mary Hesse that in order to understand complex phenomena we must create an analogy to a known entity. The positive analogy presented in this book is the comparison of the global network with other complex systems. As has been pointed out by Baruch Barzel and A.L. Barabási (2013), even among diverse networks from nature, technology, and society, topology shares universal characteristics. The same stunning commonalities can be seen for the global network of science. This tendency of real networks to share common features gives us a new handle by which to grasp the immensely complex and amazingly productive network of science.

This book suggests that understanding science communication is perhaps the best way to interject governing principles. This book builds an argument for viewing the global network as a new form of organization of science. Chapter 1 details the many changes that have occurred with the knowledge-creating system since the 1980s. Chapter 2 describes the scale and scope of the current global system with data and statistics about its scale and scope at this writing. Chapter 3 describes the communications dynamics within the global network by detailing the tenets of complexity theory to provide insights in the mechanisms driving the global system. Complex systems theory provides the framework within which to discuss the phase shift of scientific knowledge from an era of scarcity to one of abundance—a theme that will appear in the next several chapters.

Once complexity theory provides a foundation to its fuller understanding to mechanisms of change, Chap. 4 applies network structures and principles to science. Networks are the actualization of complex systems, and we can use network analysis to begin a process of understanding the structure of the global network, exploring how networking changes the system. It also allows us to visualize and measure the global network at various levels.

Chapter 5 presents the structure of the global network itself as well as the rules, norms, and practices associated with openness and abundance in the global knowledge-creating process. The creation of scientific knowledge includes not just tangible or codified output but also new practices, and even new belief systems. The rules, norms, and practices are adapting to the changing system, but some aspects of knowledge remain the same and cannot or will not change: this includes the amount of time needed to master a body of scientific work. The rules for managing networked systems differ from those that were applied under national, political systems and conditions of scarcity; we will unpack these rules in Chap. 5.

Chapter 6 presents the dynamics at the team level associated with collaboration and networked structures. Investment in science and technology is shifting. As it does, the structure of collaborative teaming is changing, too. Chapter 6 describes the emerging landscape of collaboration for the individual and the team.

Chapter 7 discusses the collaborative patterns at different geographical aggregations (local, national, regional, and global) over the most recent decade and explains why these patterns are the essence of the collaborative system and the clearest sign of a new era. Drawing upon data on global networks of collaboration, this chapter points to changes in the way scientific knowledge is created, protected, and shared at regional and national levels. The chapter also discusses networks within fields of science as well as global collaborative patterns.

Chapter 8 focuses on challenges associated with making useful knowledge truly and universally accessible. Finally, Chap. 9 discusses the policy implications, governance rules, and measures applicable to the emerging global system of science.

Columbus, OH, USA   Caroline S. Wagner

References

Adams, J. (2013). Collaborations: The 4th Age of Research. *Nature, 497*, 557–560.

Barzel, B., & Barabási, A.-L. (2013, September). Universality in Network Dynamics. *Nature Physics, 9*. https://doi.org/10.1038/nphys2741

BOAI. (2001). see: http://www.budapestopenaccessinitiative.org/

Borgman, C. L. (2010). *Scholarship in the Digital Age: Information, Infrastructure, and the Internet*. London: MIT Press.

Cetina, K. K. (2009). *Epistemic Cultures: How the Sciences Make Knowledge.* Cambridge: Harvard University Press.

Crane, D. (1972). *Invisible Colleges: Diffusion of Knowledge in Scientific Communities.* Chicago: The University of Chicago Press.

David, P. A. (2004). Understanding the Emergence of 'Open Science' Institutions: Functionalist Economics in Historical Context. *Industrial and Corporate Change, 13*(4), 571–589.

Garfield, E. (1972). Citation Analysis as a Tool in Journal Evaluation. *Science, 178,* 471–479.

Hagstrom, W. O. (1965). *The Scientific Community.* New York: Basic Books.

Hesse, M. (1963). A New Look at Scientific Explanation. *The Review of Metaphysics, 17,* 98–108.

Lichtenthaler, U., & Lichtenthaler, E. (2009). A Capability-Based Framework for Open Innovation: Complementing Absorptive Capacity. *Journal of Management Studies, 46*(8), 1315–1338. https://doi.org/10.1111/j.1467-6486.2009.00854.x.

Mann, M. E., Bradley, R. S., & Hughes, M. K. (1999). Northern Hemisphere Temperatures During the Past Millennium: Inferences, Uncertainties, and Limitations. *Geophysical Research Letters, 26*(6), 759–762.

Merton, R. K. (1957). Priorities in Scientific Discovery: A Chapter in the Sociology of Science. *American Sociological Review, 22*(6), 635–659.

Moravec, H. (1998). When Will Computer Hardware Match the Human Brain. *Journal of Evolution and Technology, 1*(1), 10.

Müller, K. H., & Riegler, A. (2014). Second-Order Science: A Vast and Largely Unexplored Science Frontier. *Constructivist Foundations, 10*(1), 7–15.

Nentwich, M., & König, R. (2012). *Science 2.0.* Frankfurt: Campus Verlag GmbH.

Padgett, J. F., & Powell, W. W. (2012). The Problem of Emergence. Chapter 1. In J. Padgett & W. Powell (Eds.), *The Emergence of Organizations and Markets* (pp. 1–29). Princeton: Princeton University Press.

Price, D. J. (1963). *Big Science, Little Science.* New York: Columbia University Press.

Thomas, L. (1978). *The Lives of a Cell: Notes of a Biology Watcher.* New York: Penguin.

Wilson, E. O. (1999). *Consilience: The Unity of Knowledge* (Vol. 31). New York: Vintage Books.

CONTENTS

ABBREVIATIONS

BRIC	Brazil, Russia, India, China (high-growth countries)
CERN	European Organization for Nuclear Research
EU	European Union
GBARD	Governmental Budget Allocations for Research and Development
GDP	Gross Domestic Product
GERD	Gross Expenditures in Research and Development
JCR	Journal Citation Report
NIS	National Innovation System
NSF	National Science Foundation (US government agency)
OECD	Organization for Economic Cooperation and Development
R&D	Research and Development
S&T	Science and Technology
UN	United Nations
UNESCO	United Nations Education, Science, and Cultural Organization
UNIDO	United Nations Industrial Development Organization

LIST OF FIGURES

List of Tables

CHAPTER 1

Science in the Age of Knowledge Abundance

In 1613, a scholar named Barnaby Rich wrote: "One of the diseases of this age is the multiplicity of books! They doth so overcharge the world that it is not able to digest the abundance of idle matter that is every day hatched and brought forth into the world." This cry could easily be uttered now, in the current age, when we feel overwhelmed by the production of information. But the sentiment of being overwhelmed by the volume of published knowledge is not new. Even the Bible warns against becoming distracted by reading too many books: "Of making many books there is no end, and much study wearies the body."[1] Yet, today, in addition to the volume of books, we have a proliferation of sources of knowledge—blogs, e-zines, proceedings, websites, audio and video presentations (some might say this is all "information" rather than knowledge, but we will discuss that)—causing us to sense that a transformation of deeper patterns is under way. The flood of information is pressing upon us in a way that is shifting how we organize ourselves as knowledge users and how we task, fund, and support the institutions designed to create and use it. Perhaps we are getting closer to the vision that Lewis Thomas (1974) had, one of interliving shoals of thought: "Later, when the time is right, there may be fusion and symbiosis among the bits...."

[1] Ecc 12:11–13 NIV.

© The Author(s) 2018
C. S. Wagner, *The Collaborative Era in Science*, Palgrave Advances
in the Economics of Innovation and Technology,
https://doi.org/10.1007/978-3-319-94986-4_1

This book addresses a specific part of the knowledge-creating process—the activities we call "science," the understanding of the natural world—and, specifically, the abundance of and access to this information. Scientific knowledge is contained in books and articles, of course, but science is also a practice conducted through research and experimentation that has spread around the world over the three centuries that constitute the years since a rational, experimental approach was designed in Western Europe by the "Invisible College" (Crane 1972). The most basic of these activities—teaching and research—are almost entirely funded by public money. Thus, it needs to be governed and accounted for to properly steward public funds. And it needs to show public benefit. This is a specific challenge of public science that creates conditions that challenge governance, as we will discuss.

Science is often referred to as a pure public good: one that is available to everyone because it is nearly impossible to exclude others and because it can be reused endlessly without depleting it. Businesses steer away from creating public goods because they cannot make a profit from something to which they cannot claim ownership. Governments generally provide for public goods and attempt to ensure the diffusion of these goods. This is truer in science than in other goods, where a large percentage of funding of basic research is the purview of governments.

Scientific knowledge is also characterized by what is sometimes called a "network effect" in that it gains value by being shared. As people grow to hold a common understanding of a natural phenomenon, such as Einstein's relativity, they can put that knowledge into use to create practices, technologies, or other new knowledge and commercial products, such as satellites. Similarly, think about the usefulness of knowing the properties of disease-causing bacteria first suggested by Lister and Koch, and tested by Pasteur. No doubt, your cupboard is filled with anti-bacterial products designed to combat the negative effects of these unseen critters. Similarly, knowledge about the need to replace electrolytes in children with diarrhea has instituted the simple, low-cost, and highly effective practice of rehydration therapy around the world, saving millions of lives. Scientific knowledge put into practice becomes common knowledge.

The sense of being overwhelmed by the flood of information is a fundamental human experience not necessarily limited to our time but tied to the limitations of our senses and life span. Who can really know all there is to understand about the universe, even as we revel in spectacular findings such as the existence of exoplanets? Scholars in nearly every age have complained that knowledge is coming at them too fast to digest. Philosopher Friedrich Hayek suggested that the "evolution of ideas has its own laws and depends very largely on developments that we cannot predict." Yet, is it completely true that the laws of the development of ideas cannot be "predicted," as Hayek suggested? The evolution of ideas has appeared in the past to be random, the result of "serendipity"—but we perceive many phenomena as random or at least as resulting from some hidden order—like the formation of beehives—until we understand it. Could it be possible to better understand the unfolding of abundant knowledge, especially scientific and technical knowledge, which has proved so useful? Could an improved understanding of the production of scientific knowledge help us improve outcomes and better manage the "data deluge," as Christine Borgman (2010) has named it? This book asks these questions and draws some lessons from the science of networks, ecosystems, and complexity itself to better understand the workings of the enterprise we call science.

Science aims to make the world a more intelligible place through observations of the many variations and differences in the natural world. This book aims to make science more intelligible to those who govern it. Science places the phenomena of the natural world within a coherent framework that makes sense in a context that transcends culture. This book explains scientific knowledge creation as networked communications that can be understood and influenced with new understandings of networks. The idea is to view science—not simply from the perspective of the record of its achievements—but as a practice of communications that has regularities that can be anticipated and, therefore, governed effectively. This exploration sheds light on the process that has been referred to as a "black box" (Rosenberg 1982) or "Pandora's Box" (Gilbert et al. 1984). By unpacking recent breakthroughs from network studies and complex systems theory, we can find new ways of understanding and perhaps influencing the "black box" of science to improve outcomes.

The convergence of many factors makes this possible. These include the demonstrated usefulness of open scientific communication, the need to solve global problems that depend on science, the information revolution, the economic growth occurring in many parts of the world, and greater political and social openness the world over. These features of the early twenty-first century contribute to deeper patterns of change for science that make it worthwhile to take time to consider the features that make it useful and how to reimagine it for a global era. Many parts of scientific practices continue to operate as they have done for centuries—white-coated men and women still swirl chemicals in a beaker—but other parts of its practice are changing, such as a global network of seismometers that record the occurrence of hundreds of earthquakes every day across the globe. An enhanced understanding and better stewardship of the processes can also improve the ways we create and use it. After all, we "own" it—we pay for it and expect science to provide knowledge that we need. In many ways, society is looking to science to address the most pressing and significant problems of climate change, energy needs, food availability and security, water usage, and universal education. It is worth taking time to consider the underlying forces of organization, use, and benefits of knowledge. This book addresses these questions.

The practice called "science" has undergone many shifts and changes over time. The history of science is fascinating in itself. Even the word "science" was only applied in the eighteenth century to the practice of creating reproducible knowledge about the natural world. This book focuses on the future, however, and in particular on two aspects of science operating at two frontiers: one, science at the international level where the practice of scientific collaboration has been growing rapidly; and two, at the interdisciplinary frontier, where it is possible to see a good deal of what appears to be growth in the practice of "interdisciplinary" science. These areas of scientific practice have not been well studied, even though they are growing rapidly. These two areas, which also intersect and overlap, are more likely than other aspects of science to operate as networks for reasons that we will discuss in the book. For reasons that will become clear as we lay out their structure and rules, networks are what you would expect to find here because of the emergence and complexity of both international (or global) science and interdisciplinary research.

Characterizing science as a network opens up new and effective ways to describe the dynamic processes. Network science has come into its own over the past two decade. An enhanced understanding of networks can help those of us who support, use, and manage science to improve the creation and dissemination of "useful knowledge" as Joel Mokyr (2002) calls it in his book, *The Gifts of Athena*—knowledge that is "accumulated when people observe natural phenomena in their environment and try to establish regularities and patterns in them" (p. 3). Mokyr's definition is broader than the classical definition of science as verifiable knowledge about the natural world, in that "useful knowledge" can include indigenous knowledge and practices that are often considered to be outside the scope of science. Thus, in Mokyr's sense, useful (he calls it "propositional") knowledge includes informal ways of knowing about nature—close to what the physicist-turned-philosopher, Michael Polanyi, called "tacit knowledge"—and can inclusively extend to the knowledge and practice of the artisan or the medicine woman. Local, indigenous, or folk knowledge shares the characteristics of being verifiable and reproducible, but it is not often considered to be in the realm of reproducible science even though it is useful.

The aspects of inquiry that we call science have both common and variable aspects. The common ones surround the scientific method of experimentation, reproducible results, and the communication of findings. Variables are the object of study, how it is done, who pays for it, and how it is shared. These variables are often determined by social or political factors, and are therefore subject to change. These are the factors that motivate this particular inquiry. The research for the book was motivated by the observation that, for many of the policymakers, philanthropists, and investors seeking to include more people into the practice and understanding of science, the shifts in the social and political aspects of science lead to frustration. For example, the European Commission has invested billions of euros in research and development, yet their scientific outputs still lag behind that of the United States in many indicators of "scientific quality." The World Bank has provided millions of dollars in grants and loans to help poor countries develop useful knowledge, but these nations remain poor. The United Nations has encouraged science and technology investments to help the world reach the Millennium Development Goals (MDGs), yet this has not resulted in propelling the poorest peoples to achieve the hoped-for outcomes. This is not to say that there has not been progress but to note that inequality remains a problem. Science does not, in itself, reduce inequality.

Science cannot solve all these problems. Yet, among those people who seek to apply scientific solutions to these problems, we need a solid understanding of how science operates, a feature that changes to some extent with the social and political structure. Modern science has transitioned from a collection of largely independent experimentalists in the seventeenth century, to laboratory-based research supported by governments and donors in the eighteenth century. As science proved its worth, it became a profession for which young people were trained in universities, and governments increased support for it during the nineteenth century. In the twentieth century, science became largely captured by nations, and its practice was highly tied to national prestige and the service of war and economic growth. The vestiges of each of these previous eras remain, but the practice of science is undergoing a fourth metamorphosis in the twentieth-first century, shifting toward global networks that operate beyond the silos and constraints of the institutions and disciplines that defined it in earlier days.

This shift from national and disciplinary identities to locally useful, globally connected knowledge networks changes the way we organize and account for science. If the "measure" of science were inverted from a national competition toward one of measuring increased welfare (due to integration of useful knowledge), would the European Commission still view itself as suffering a "European paradox" when compared to the United States? If local knowledge—even know-how emanating from the medicine woman and local artisans—were to be considered as valuable as scientific publications, would the poorest countries still be considered to be outside the realm of the "knowledge society"? If the United Nations' well-meaning MDGs were met with an integrative process linking scientific knowledge to the underlying goals, would the world be closer to closing gaps in clean water, maternal health, or clean food? Right now, we can only answer, "perhaps." But it is the competitive and exclusive face of science that has left some feeling "behind." A change in the way we think about science could change this for the better.

With improved insights into the knowledge-creating process—specifically, to shift away from science as a "national asset"—to view knowledge as a networked resource operating at local, regional, and global levels—it is possible to create new and more effective policies and governance tools to extend the benefits of science to a broader user group. The insights presented here are designed to help to improve the experimental processes and move knowledge more readily toward useful applications, increasing efficiencies, and extending science to include new participants.

The Study and Understanding of Science

The nature of science and the intricacies of its operations have been a subject of inquiry since the time of the early Greeks. Many writers have tackled the question of "what is scientific knowledge?" Many earlier efforts have emerged from within sociology and philosophy by scholars concerned with broad cultural implications of the impacts of scientific knowledge. Science has been seen as a blessing and a curse. It has been seen as a noble endeavor and a foolish practice. It has been held up as the pinnacle of human achievement, and it has been forced underground by political regimes threatened by it. The practice of science has shifted in scale, scope, or style along with the changes in social and information infrastructures.

Among the ideas about knowledge creation in science, perhaps the best known is *The Structure of Scientific Revolutions* in which Thomas Kuhn (2012) described his view of how science advances through a process of shifts between normal and revolutionary science among communities of scientists. Kuhn stated that scientists spend most (if not all) of their careers in a process of problem-solving. Their problem-solving is pursued with great tenacity because the previous experiments created accepted facts about an established "paradigm." The paradigm tends to generate confidence that the approach being taken within that field guarantees a solution to a puzzle—this can be called "scientific theory." Kuhn calls this process "normal science," and indeed the bulk of scientific practice still looks quite similar to what Kuhn called "normal science."

Kuhn went on to describe how science changes in his view: as a paradigm is stretched to its limits, anomalies occur—strange findings that fail to meet expected outcomes detailed by the existing paradigm. The accepted theory cannot fully account for or explain an observed phenomenon or data. This is the moment that Isaac Asimov calls the greater moment in science than "Eureka!" but one where someone says "that's funny." An example of an anomaly was made public in the summer of 2011. Scientists in Europe, testing the speed of light, could not account for an anomaly. Physicists at the Gran Sasso Laboratory, Italy, had data showing certain neutrinos moving from a site in Switzerland to the Italian site at speeds that appeared to be faster than the speed of light. Obviously this challenges accepted laws of physics. The teams considered and debated for some time, but in the end, they decided to take the story to the broader scientific community by using the media—the wisdom of the crowd, shall we say. The questions raised caught international attention beyond the scientific community, and great speculation ensued about the implications of a change in the basic laws of matter. A few months later, it was discovered that a technical problem with an optical fiber cable had provided false data, so the anomaly was settled: the neutrinos did not travel faster than light. Einstein could rest easy, although not the heads of the two labs involved, who were fired.

Some anomalies may be dismissed or proved to be errors in observation, as with the Italian "faster than the speed of light" story. Others may merely require small adjustments to the current paradigm or theories in ways that are clarified in due course. Some anomalies resolve themselves spontaneously, having increased the available depth of insight along the way. Some anomalies are created when new equipment makes additional testing possible. But no matter how great or numerous the anomalies that persist, according to Kuhn, the practicing scientists do not lose faith in the established processes to address them as long as no credible alternative or reproducible experiment with solid data is available to say otherwise. This part of the story of science seems to resonate with current experience, and many observers would recognize this as "the way science is done."

In any community of scientists, Kuhn stated, there are some individuals who are bolder than others—these might be considered as the scientific equivalent of a business entrepreneur. These scientists, judging that a crisis is brewing, embark on what Kuhn calls revolutionary science, asking questions and exploring alternatives to the firmly established assumptions. Occasionally, this process generates a rival theory to the established frameworks of thought—a new theory that can be tested and discussed openly among the relevant communities. The new candidate paradigm usually appears to be accompanied by numerous anomalies of its own, perhaps because it is still so new and incomplete, or perhaps because it is wrong. The majority of the scientific community will resist any change to theory, and, Kuhn emphasizes, so they should. Science requires a strong conservative core of skeptics who hold onto existing understandings until the new paradigm has been robustly defended or proven by many to be wrongly assumed.

"Crises" (as Kuhn called them) in science are not always resolved with the development of a new theory or paradigm. There are many examples in the history of science in which confidence in the established paradigm or theory was eventually vindicated. Whether the anomalies put forth as a reason to question the existing theory are resolved in favor of a new paradigm or not is almost impossible to predict. Sometimes the equipment to test a new theory simply does not yet exist, and facts cannot be recalibrated until some future time. But the process of scientific experimentation welcomes these excursions into exploratory research. Those scientists who possess risk-taking characteristics and are willing to entertain a theory's potential may be the first to shift in favor of the challenging paradigm, and be richly rewarded for doing so. But this can also be the domain of the foolhardy. The only way to tell the difference is by testing and retesting the phenomenon with accepted processes. If anomalies remain unresolved, there typically follows a period in which there are adherents of two or more paradigms in conflict with one another. In time, if the challenging paradigm is solidified and unified, it will replace the old paradigm, and a *paradigm shift* will have occurred; a new theory is born.

Kuhn's theory of shifts in the structure of scientific understanding fits science of the twentieth century so well that his ideas came to dominate most of the descriptions of the scientific process. The concept of anomalies, crises, and paradigm shifts are particularly apt when discussing disciplinary-based sciences of what was affectionately called "the ivory tower"—those academic institutions that housed scientists in departments and laboratories, places that served as bulwarks against capricious changes in science. Within the walls of these institutions was the codified word (books, journals, notes) that constituted science and where young scientists came to be trained to continue carrying the core of knowledge and the rules for changing it. The "knowledge" of science was difficult to access, and entry into the fraternity of scientists was carefully controlled by a small group of elites. Full brotherhood required spending years learning a special language and practice that was understood by only a few others similarly initiated into the disciplines that constituted science in the nineteenth and twentieth centuries.

MODES AND COMMUNITIES

At the end of the twentieth century, another approach to understanding science was added to the notion of paradigm shifts, perhaps because the social and political practice of science itself was changing. Some academic disciplines continued to operate in ways consistent with Kuhn's ideas, but another approach was offered called "Mode 2" knowledge (Gibbons et al. 1994). Recognizing that Mode 1 (academic, disciplinary science) continued to operate, a team of European scholars asserted that the knowledge-creating process was undergoing a dramatic shift, both in the institutional context of knowledge production and in the kind of knowledge being produced. Led by British sociologist, Michael Gibbons, the team argued for understanding science as having two "modes" that exist in parallel. Mode 2 was the term applied to knowledge processes emerging over and above Mode 1 (Kuhnian "normal science") in response to a number of external factors. Mode 1 was identified as "traditional knowledge" generated within a specific disciplinary, cognitive, and primarily academic context. Mode 2 was knowledge generated outside academic institutions in broader, trans-disciplinary social and economic contexts, often to meet a specific need or goal. The authors asserted that the rise of Mode 2 did not negate or obviate the existence of Mode 1 operations, but they asserted that much more science—and more interesting science—was being done according to the Mode 2 model.

In the view of the Mode 2 proponents, the new methods of knowledge creation may have arisen because of what they termed the dramatic expansion of higher education opportunities, ones that created what they called a "surplus of highly skilled graduates"—people who could not be absorbed into traditional academic settings simply because there were not enough jobs in the academy. Since these newly trained scientists could not find positions in traditional domains of science, the authors contended that they applied their knowledge to needs found in private industries and laboratories, or in some cases in their own start-up, science-based business consultancies, or think-tanks like the RAND Corporation or the Wellcome Trust. The consequence of this spillover from academe into a larger business and social context was the proliferation of multiple sites of knowledge production, with universities no longer holding a monopoly on the certification of valid knowledge.

Mode 2 proponents described certain key changes that occurred as a consequence of the emergence of multiple sites of non-academic knowledge production, the most notable of which was that knowledge was being produced in the context of applications related to problem-solving approaches rather than as basic puzzle solving. Donald Stokes (2011) made a similar observation in 1997 when he suggested that even basic scientists could name practical applications for their research. (In academic circles—at least in the twentieth century—this "application" focus would be viewed as adulterated science not worthy of the academy.) In contrast, knowledge produced in Mode 2 could be characterized as "trans-disciplinarity" or outside of the disciplinary structure that existed within academe. In Mode 2, solutions to problems to which science could be applied were not constrained by disciplinary limits. As they saw it, Mode 2 processes created knowledge by using a complex network of linkages drawn from wherever knowledge could be useful. Mode 2 was also characterized by "social accountability" and "reflexivity" (a process of feedback loops that consider the meaning of the knowledge created and how well it is helping to solve specific problems) that further differentiated it from Mode 1 or academic science. The Mode 2 team contended that, contrary to criticisms that might be hurled by those in the Ivory Tower, by working in the context of applications and problem-solving, these socially aware scientists and engineers became more sensitive to the broader implications and impact of their action, making them sensitive to the public concerns about the environment or other social issues. Ways in which science could contribute to competitiveness in the marketplace, cost-effectiveness, ethical applications of knowledge began to become as important to the new generation as peer review had been to the Ivory Tower academics.

Karin Knorr Cetina (2009) describes similar processes and changes in the knowledge-creating process in her 1999 book *Epistemic Communities: How the Sciences Make Knowledge*. Her efforts included time spent in laboratories examining the work of scientists, seeking, as she says, "to open up the black box that constitutes scientific inquiry and make sense of the various activities" in science by understanding practice and culture related to knowledge. Cetina argued that knowledge was becoming the productive force within economies, rising in stature to equal land, labor, and capital—those productive forces that were counted as the essential ingredients for economic growth. The addition of "knowledge" as a primary productive force should include the role of information infrastructures and the associated changes in economic and social organization that result from them. Cetina asserted that a "knowledge society" is not simply a society with "more knowledge" but also a society "permeated with knowledge settings," where many, if not most, of the features of society are driven by access to, use and production of, knowledge. Structuring social change or investment should consider the need to include a knowledge infrastructure and the features and functions to disseminate knowledge to those who can use it, according to Cetina.

Cetina asserted that epistemic cultures should be designed to capture and make useful what she calls "these interiorised processes of knowledge creation" within the mind and physical environment of the researcher (p. 363). She refers to "those sets of practices, arrangements and mechanisms bound together by necessity, affinity and historical coincidence which, in a given area of professional expertise," make up how we know what we know in science. But these insights cannot and should not stay within the laboratory if society is going to become a fully knowledge-driven entity. She chose the term "epistemic" rather than simply "knowledge" in order to encompass this broader concept of understanding combined with and practice. In contrast to Kuhn, Cetina's focus is more on *action* than on knowing—how does the environment (such as the laboratory, or the symbols one uses) influence the outcomes? How can they be put to use? Tying them more closely to social needs is one link that should be made, she says.

Cetina considers within the purview of her inquiries the laboratory and its equipment, scientific measurement processes, the objects of research, and the language of interaction. Cetina presents the knowledge-creating process as driven by the actors within the sciences, manipulating these various factors in their environment and using them to create meaning and context. The epistemic communities create knowledge through interaction, and they become skilled at this process. However, in her view, the epistemic community should not be tasked with the extra job of disseminating knowledge, since this is not what they are good at doing. This process falls to social and political actors. Knowledge of the internal workings of epistemic communities can help those actors who seek to translate science into social goods. Cetina's descriptions of communities do not conflict with or confirm Kuhn's theories nor do they challenge Mode 2 of operation. Her ideas can be seen as adding to rather than replacing these concepts of science. She sought to shed light on how scientists operate.

Where Karin Knorr Cetina focuses on the scientist within the laboratory, Ikujiro Nonaka and his colleagues examine the process of knowledge creation at a systems level, offering a wholly different theory of knowledge creation. In Nonaka's view, knowledge creation is a process of synthesizing ideas occurring within a communications framework. Nonaka writes mainly for a business audience, and his ideas are put in the context of the firm, but many of the concepts are applicable to public-goods science. Writing in the early part of the twenty-first century, Nonaka's view was that "we are still far from understanding the process in which an organization creates and utilizes knowledge. We need a new knowledge-based theory that differs 'in some fundamental way' from the existing economics and organizational theory" (p. 2). Nonaka asserts that many scholars and practitioners, even while they are immersed within the knowledge-creating process, fail to understand the essentials of knowledge creation. Knowledge creation is a dialogue—a process of exchange of ideas and information from which a common concept emerges. Nonaka called this a "dialectical process" that occurs through dynamic interactions among individuals, the organization, and the environment, where the relationship of actors and structures is a mutual one. Nonaka shares the view of the importance of communications with other scholars such as German sociologists Niklas Luhmann or Max Weber, who viewed knowledge creation as a dialogue—a process of exchange of ideas and information from which a common concept emerges.

Nonaka dubbed the landscape of knowledge creation as "ba"—a shared space where knowledge is created in a spiral of events that go through "seemingly antithetical concepts such as order and chaos, micro and macro, part and whole, mind and body, tacit and explicit, self and other, deduction and induction, and creativity and efficiency."[2] Nonaka argues that the key to understanding the knowledge-creating process is "dialectic thinking and acting"—processes of communicating and collaborating. In his view, this process transcends and synthesizes contradictions that appear within the communications processes. The contradictions become a pathway to grow into a new, shared understanding which is not "compromise," but represents a new understanding. The integration of diverse and sometimes opposing aspects of a problem (a process similar to Kuhn's anomalies) takes place through a dynamic process of dialogue and practice.

Nonaka's view of knowledge creation resonates with the ideas of Niklas Luhmann (1982) especially, for whom communications were the organizing force for social structure for knowledge creation. Luhmann argued that complexity is the mark of modern society, containing within it many emergent properties—a concept we will spend more time on in Chap. 2. Like Nonaka, Luhmann does not start with the individual when considering how knowledge is created; rather, both these thinkers consider the role of society and communications processes as reflecting back onto the individual a structure within which he or she already operates. These assertions that knowledge creation has a structured environment are also not new in themselves, since other sociologists have described social structure as important for knowledge creation and retention, but they differ from others in seeing the structure as highly determinant of the opportunities available to knowledge creators. As Nonaka points out, the participants in the knowledge-creating process "coexist with the environment because they are subject to environmental influence as much as the environment is influenced by the[m]."[3]

The idea that the structured and rational processes of science could become the subject of scientific study is a concept that was not looked upon kindly in the earliest days of science. In part it was questioned

---

[2] Nonaka and Toyama (2003, p. 2).

[3] Nonaka further notes that this conceptualization of the interdependent connection between the entities and structure is similar to Giddens' structuration theory (Anthony Giddens 1984). The existing theories that deal with a static status of an organization at one point in time cannot deal with dynamic processes, but that is the benefit of using a network approach.

because of its inherent interdisciplinarity—the science of studying science occupies an intellectual space that is "between" many other fields such as sociology, philosophy, economics, politics, and, more recently, analytics. As the "science of science" has grown, and a distinct subfield of science policy has emerged as its own discipline, both subfields have adopted many of the earlier models of science. Among the challenges facing those who take up this baton is the one of explaining science to non-scientists, especially to those who are responsible for promoting and using it to reach social and economic goals. As the attributes of science change, we are compelled to analyze once again the context within which it operates. The ideas about how to understand, interpret, and make the best use of science must change with the changing social and political contexts within which it operates, and the next chapter aims to add a description to those that come before, to adapt to the current landscape.

CONTINUING A DISCUSSION ON KNOWLEDGE CREATION IN SCIENCE

This topic is not new: many people have written about knowledge creation in science. Many thinkers of the last century focused attention on the role of the scientist in society as well as the influence of scientific thought and outputs on society. It continues to fascinate. It is challenging because of its inherent complexity. It can be viewed from many angles. As one theorist wrote in 1971: "The idea that the development of science can be analyzed at all effectively, apart from the concrete research of scientists, is said to have proven false. The study of science, after all, has recently focused on its product, scientific knowledge, rather than simply with those individuals who occupy the social positions of 'scientist'." This quote reveals a point of view that has dominated science studies in the post-war era, that of studying the published works of science. Less attention overall has been given—at least in recent decades—to the scientists themselves. The study of the scientist on one hand and the study of the outcomes of science in written form on the other hand can be seen has trending along two lines, one examining the motivation of the individual scientist in works such as Bruno Latour's *Science in Action* (1987), and the other studying the collected works of scientists in quantitative assessments called bibliometrics and scientometrics set in motion by Derek deSolla Price in *Big Science, Little Science* (1963). The two traditions very much distinguish between the behavior of the scientist as "practitioner" and the outputs of science as "knowledge," with different underlying assumptions about the role of the individual in influencing science and the role of collected knowledge.

Along with a focus on the individual or a bibliometric strand, a third thread is woven into science studies—the study of scientific process and "knowledge as a system," which follows more along the lines of systems theory, asserting that knowledge dynamics can be understood as a complex system of communications. This strand examines the scientific system as existing over and above the role of the individual scientist. These studies led by Loet Leydesdorff (2001) have concepts in common with earlier work by Talcott Parsons (1961) and Edward Shils (1975), sociologists who sought to define a more general theory of action by distinguishing categories and relationships, although they focused on groups but not exclusively on science. Edward Shils applied structural ideas to personal knowledge. Shils suggested that values and beliefs form a "center" within a community that offers a touchpoint around which other social organization can take place. Several analysts including Sheila Jasanoff (2004) have carried on with research around this theme, offering a social-constructionist model of science and its interaction with and impact on society.

This book attempts a convergence of all three threads by considering each one as representing a level in a complex system. Since existing concepts can only go so far to help us to understand the current world which manifests a complex social reality, the convergence of disciplines, the inversion of public and private concerns, the reorganizing of political alliances, and the development of a global mentality, a new model is needed. The reorganization of the sciences—with the emergence of global science as the most visible of these facets—is a part of a system as a whole reorganizing itself around dynamics of communications processes. No longer do we use a rhetoric where we describe a timeless, machine-like universe of Isaac Newton: this notion has been replaced by a universe of disorder, non-linear complexity, and unpredictability, where paradigms can be ephemeral and shifts occur with increasing frequency. Some thinkers have grabbed onto the chaotic and random riffs as evidence of a universe that hurtles toward disorder (and perhaps disaster). These social interpretations of science often misunderstand and misrepresent the meaning of science: a concept that will be discussed in the last chapter. Complexity theory as a way to understand science now offers a more meaningful and helpful approach to understanding the role of science in society.

In the midst of the shifting roles of science and scientists, the practice of science continues to operate with a good deal of social support. In fact, as many other institutions of society are up-ended (journalism, medicine), science as a social institution appears to grow in stature and trust by the public. Whether this is good or bad remains to be seen, but at the very least, we need new ways to discuss what we think of as science and how it serves society. No longer do we view science as a private activity of few wealthy or eccentric people with obscure goals. In fact, Paula Stephan (2012) has pointed out that the public appears to have a growing sense ownership of science. Thus, the dynamics of science can be viewed as collective action that operates within a commons—a place of collectively maintained and shared resources that are provided by and for the general public and their welfare. A large number of conditions influence the prospects and accountability for this collective action, and finding appropriate ways to discuss it becomes an imperative. We can see that taking these conditions apart to examine and manage each one separately cannot and does not reveal the dynamics at the systems level. This is the core of the problem facing science today—we have little understanding of the systemic nature of science.

REFERENCES

Borgman, C. L. (2010). *Scholarship in the Digital Age: Information, Infrastructure, and the Internet.* London: MIT Press.

Cetina, K. K. (2009). *Epistemic Cultures: How the Sciences Make Knowledge.* Cambridge: Harvard University Press.

Crane, D. (1972). *Invisible Colleges: Diffusion of Knowledge in Scientific Communities.* Chicago: University of Chicago Press.

Gibbons, M., Limoges, C., Nowotny, H., Schwartzman, S., Scott, P., & Trow, M. (1994). *The New Production of Knowledge.* London: Sage.

Giddens, A. (1984). *The Construction of Society.* Cambridge: Polity.

Gilbert, G. N., Gilbert, N., & Mulkay, M. (1984). Opening Pandora's Box: A Sociological Analysis of Scientists' Discourse. CUP Archive.

Jasanoff, S. (Ed.). (2004). *States of Knowledge: The Co-production of Science and the Social Order.* London: Routledge.

Kuhn, T. S. (2012). *The Structure of Scientific Revolutions.* Chicago: University of Chicago Press.

Latour, B. (1987). *Science in Action: How to Follow Scientists and Engineers Through Society.* Cambridge: Harvard University Press.

Leydesdorff, L. (2001). *The Challenge of Scientometrics: The Development, Measurement, and Self-organization of Scientific Communications.* Boca Raton: Universal-Publishers.

Luhmann, N. (1982). *The World Society as a Social System*. Taylor & Francis.

Mokyr, J. (2002). *The Gifts of Athena: Historical Origins of the Knowledge Economy*. Princeton: Princeton University Press.

Nonaka, I., & Toyama, R. (2003). The Knowledge-Creating Theory Revisited: Knowledge Creation as a Synthesizing Process. *Knowledge Management Research & Practice, 1*(1), 2–10.

Parsons, T. (1961). Some Considerations on the Theory of Social Change. *Rural Sociology, 26*(3), 219.

Price, D. J. (1963). *Big Science, Little Science*. New York: Columbia University Press.

Rosenberg, N. (1982). *Inside the Black Box: Technology and Economics*. Cambridge: Cambridge University Press.

Shils, E. (1975). *Center and Periphery* (p. 3). Chicago: University of Chicago Press.

Stephan, P. E. (2012). *How Economics Shapes Science* (Vol. 1). Cambridge, MA: Harvard University Press.

Stokes, D. E. (2011). *Pasteur's Quadrant: Basic Science and Technological Innovation*. Washington, DC: Brookings Institution Press.

Thomas, L. (1974). *Lives of a Cell: Notes of a Biology Watcher*. New York: Penguin Press.

CHAPTER 2

The Scale and Scope of Global Science

It is now widely accepted that science and technology underlie much of the economic growth that has been a highlight of human social development for over 300 years. Science has grown along with social welfare, economy, and the application of human imagination in a way envisioned by Ada Lovelace (Toole 1992): "Those who have learned to walk on the threshold of the unknown worlds, by means of what are commonly termed par excellence the exact sciences, may then, with the fair white wings of imagination, hope to soar further into the unexplored amidst which we live." The only way that others know you have soared further is because you communicate it. If your findings are to endure, you must publish them. The most tangible parts of scientific growth have been the exponential rise in the stock of scientific knowledge (articles, journals, newsletters, patents, and books) over the past three centuries, along with a rise in the number of people working in science and technology-related fields. It is hard not to notice that the growth of scientific output has occurred in line with the growth of the size and increasing success of the human race as a species—and also with the impact of that species on the globe, but a direct connection between science and species success is hard to prove.

A number of scholars have studied the rapid rise in the body of scientific scholarship, whose growth, amazingly, can be shown to conform to mathematical laws of growth found in other parts of nature—ones related to exponential growth. For established scientific powers such as the

© The Author(s) 2018
C. S. Wagner, *The Collaborative Era in Science*, Palgrave Advances
in the Economics of Innovation and Technology,
https://doi.org/10.1007/978-3-319-94986-4_2

United States, Japan, and European Union countries, total national research output measured in scientific papers more than doubled over 30 years, between 1980 and 2010. Domestic output (with an author or authors from a single country) has increased by about 50%—while international collaborations have accounted for a vast percentage of the increase. After the mid-1990s, the domestic research output of the United Kingdom, Germany, and France leveled off, while international collaboration in these countries increased more than tenfold. Papers that are co-authored now claim a higher share of all papers, and papers that are internationally co-authored grew even faster as a share.

Jonathan Adams (2013) has called this the "fourth age of research," but it might be more revealing to call this "the collaborative era" as this is the way we are creating and sharing knowledge—in a collective enterprise. Change is being led by the scale and scope of the outputs of science and technology as well as the infiltration of scientific or technical thinking in many aspects of life where they might once have been on the periphery or not involved at all. Collaboration and openness have had a significant impact on the organization of science and on the patterns and flows of knowledge in research, development, and innovation.

It is arguable how much of a change the current era represents from past experience, especially in a system that is always changing. In the late 1930s, the eminent scientist J.D. Bernal (1939) wrote of scientific organization that, "At present the organization of science is in a transitional stage of development; it is passing from the period when it represented a sum of individual efforts to one when it advances by conscious team-work, through which the contributions of the individual scientists are absorbed in the general result" (p. 103). On one hand, we may be overstating the change represented by the rise of collaboration and teaming, since it may be that the change is in the willingness to list co-authors. J.D. Bernal made clear that science had a collective character, even if the practice at the time was for one person to rise up to claim credit. Consider the case of the relatively unknown engineer, Oliver Heaviside, who interpreted pages of equations on electromagnetism made by James Clerk Maxwell. After studying the notes, Heaviside produced a brief, elegant set of four equations on electromagnetism that derived from and ultimately became attributed to Maxwell. This was a relatively common practice, and collaborators did not receive credit, especially if they were paid assistants.

This book makes a different argument about the change as one being from a system operating under rules of scarcity to one operating under rules of abundance. The rules of abundance clearly change the dynamics and incentives for participants, from closed to open, from protective to sharing, from individualistic to collective, from monetary to "useful." Each of these changes also alters the governance requirements. The next section discusses the scope of the global network.

From the Individual to Collective

The history of science is filled with contributions made by a group (or at least consultations) that is then named for a single person. It is also with the co-occurrence of inventions and developments that spontaneously pop up at the same time in two different places. Niels Bohr contributed to the general theory of relativity, even though it is Einstein who gets the credit. Perhaps we are simply more egalitarian now and therefore more willing to share the credit with co-authors, since many more articles are co-authored now than ever before in science. It is a rare scientist who labored in isolation.

The white-coated scientist may still labor on in some places, but the system within which he or she works is changing rapidly. The rules and incentives that worked in the past are either defunct or no longer operate in the same ways. The changing system is altering the requirements for the support systems around science, including such fundamental features as funding, patenting, laboratory space and organization, publishing, and knowledge sharing. The transformational changes in the knowledge system introduce new constraints and incentives built to encourage or control diffusion. These include invention disclosures and rights to a monopoly, imposition of export controls, and use of trade secrets. All these features of the system are being put to the test, and many are failing to show they can continue to offer the benefits they once did. Rather than being random changes, however, these features are what one would expect as a new communications ecosystem emerges.

Box 2.1 Organizations Supporting International Cooperation in Science

A wide range of organizations support, fund, advocate for, study, or aid international cooperation in science. The types of organizations include national governmental such science or technology ministries—some of which directly fund non-national researchers,

(*continued*)

Box 2.1 (continued)

and some of which do not. Intergovernmental groups such as the European Commission and ASEAN provide funds to international cooperative projects. Intergovernmental organizations such as United Nations Educational, Scientific, and Cultural Organization (UNESCO) and the International Council for Science (ICSU) provide informational, material, and, rarely, financial support for cooperation. Intergovernmental support groups, such as Organization for Economic Cooperation and Development (OECD) and the World Bank, collect and publish data about national scientific spending and output, and evaluate national and international efforts.

Non-governmental scientific societies and professional membership societies promote, coordinate, and sometimes fund research. They often play a role in standardizing research terms and processes, and they provide an informational role of convening people, planning, and exchanging and diffusing knowledge. Members often provide funds for these kinds of groups. Examples include the International Union of Geodesy and Geophysics, the International Union of Soil Sciences, or the patient-driven European Organisation for Rare Diseases (EURORDIS).

National non-governmental organizations, in some cases, play a role at the international level, including the American Association for the Advancement of Science and the Japan Society for the Promotion of Science, by convening researchers to discuss future directions for research and publishing information on scientific developments.

Academies of science, engineering, and/or medicine can include governmental, private, and educational members among their members. These groups often promote, study, evaluate, and support international cooperation in science. Notable among these has been the Third World Academy of Science that has actively sought to engage developing country researchers in global research opportunities. Similarly, the Interacademy Council has coordinated international cooperation, as well as reported on it, for more than a decade. In a model similar to the French Centre National de la Recherche Scientifique (CNRS), the Chinese Academy of Sciences is notable for its very large system, maintaining laboratories and employing researchers as well as tracking and monitoring research and development (R&D) within China.

(*continued*)

Box 2.1 (continued)

Philanthropic groups—both governmentally sponsored and privately supported—work to provide material and financial support to cooperative research, and to aid in training and convening researchers to be part of global cooperation. Governmental philanthropies that provide support for science include the Swedish International Development Cooperation Agency (SIDA) and Canada's International Development Research Centre (IDRC). Privately sponsored philanthropic groups funding international cooperation in science include the Wellcome Trust or the Rockefeller Foundation.

SIZE OF THE SYSTEM

It helps to put down a marker as to the size of the system when this work is published. The system of science continues to grow at all levels. The number of countries contributing more than 1% of their gross domestic product (GDP) to research and development (R&D) has increased significantly over the past three decades. In 1981, the average gross domestic expenditure (GERD) on R&D for OECD countries and observers was 1.59% of GDP, with 13 of 20 countries reporting more than 1% of GDP dedicated to R&D. By 2015, the average R&D/GERD had risen to 2.38 for 34 countries reporting, with 26 of these spending greater than 1% on R&D/GERD. Comparisons across countries are complicated by diverse budgeting practices, but OECD shows that many advanced nations spend between 2% and 3% of GDP on R&D. Total R&D spending (public and private) topped $1.5 trillion in 2015.

The estimated number of researchers has also grown considerably since 1981. While methods have improved toward a standard set as reported by the OECD, the numbers for other countries (non-OECD) are only loosely comparable. According to UNESCO (2010) counting OECD and beyond, about 7 million R&D practitioners were active. OECD also collects data from member and observer countries and reports an estimated increase in the number of researchers in its member countries from 3.1 million in 2000 to 4.4 million in 2012. (This number included Russia and China, but not India or Brazil.) Exceptions to the overall worldwide trend of increasing numbers of researchers included Japan (which experienced little change) and Russia (which experienced a decline; see also Gokhberg and Nekipelova 2002, referenced in NSB 2016). Using World

Bank data (Researchers per million, 2014), the estimate of the number of researchers working within 63 countries in 2014 at 7.2 million, a number which is close to the UNESCO estimate.

The number of journals publishing the results of worldwide scientific research is even more difficult to count than money or practitioners. Various attempts have been made to measure the scope and trajectory of scientific publishing. The literature on the growth of science has counted the number of articles and the proliferation of journals and other venues (such as online journals) where science is published. The rise of new venues of publication—such as arXiv, the Public Library of Science (PLoS), and Scientific Commons—complicates the counting and categorization processes. Web of Science and Scopus, which, for all their strengths, are limited by design in representing the scope of scientific output (Larsen and Von Ins 2010; Harnad 1997; Borgman et al. 2007). They report only a percentage of all publications. Jinha (2010) reported that the number of articles topped 840,000 in 1990. The rate of increase each year is 0.325 (Mabe 2003).

How Many Journals Are Actively Published?

Figure 2.1 shows the slow beginning of the number of scientific articles, as estimated by A. Jinha (2010).[1] Over time, the production of scientific articles grew rapidly, taking a sharp upward turn in the 1960s, to grow at an exponential rate, as shown in the figure.

The truly innovative contribution of the seventeenth-century "invisible college" was to affirm the practice of open sharing of the results of experimentation. Others should be able to reproduce, validate, or question the results, they asserted. This was in contrast to the practices of alchemists, who held onto the secrets of their processes. Prior to the organization of scientific societies in the seventeenth century, experiments were kept secret (often, the goal was to get rich by hoping to turn lead into gold). Experimentation was the realm of the magician, and knowledge was the purview of the priestly or scholarly classes—not to be widely shared. The social concept of a *knowledge commons*, or knowledge as a public good,

---

[1] An early example of a science abstracting service is the collaboration between the Royal Society of London and the Institution of Electrical Engineers. In 1898, the two groups jointly published *Science Abstracts* compiling 110 abstracts of articles written by a "Who's Who" of eminent European scientists. See the Institution of Engineering and Technology history of Inspec, www.theiet.org

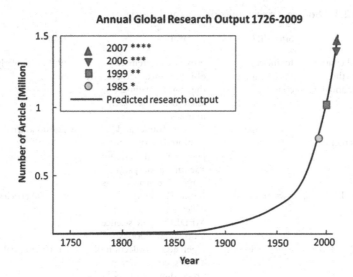

Fig. 2.1   Research outputs in numbers of articles. (Source: Jinha 2010)

does not emerge for several centuries. This concept of a knowledge commons develops slowly and unevenly. (Even Vannevar Bush (1945) in his treatise, Science: The Endless Frontier, viewed research as within the purview of a small, studied elite.) The idea of a global public good emerged much more quickly as articles and data have been digitized and made electronically available.

Table 2.1 presents various attempts at journal counts in the mid-2000s. Derek de Solla Price (1961) found that the number of scientific publications has been growing exponentially for over three centuries; he estimated about 30,000 journals in the early 1960s. Michael Mabe (2003), conducting a similar study in the early 2000s, found that the number of journals being added year-on-year has been remarkably consistent with average rates of 3.46% from 1800 to the present day. Mabe used Ulrich's Periodicals Directory to derive a count of scientific journals as of 2000: he counted 14,694 active, academic/scholarly, refereed journals.[2] Using different search criteria, sociologist Ina Jinha estimated 57,400 referred journal titles worldwide. Many of the differences can be attributed to

---

[2] Mabe notes that Meadows estimated as many as 71,000 journal titles in 1987 using a broader search string.

Table 2.1   Sources of scientific journals and articles

Source	Site/URL	Description	Records
Journal citation report (the Netherlands)	Jcr.incites. thomsonreuters. com	provides a systematic, objective means to evaluate the world's leading scientific and scholarly journals	>29,000 scholarly journals
Scopus (Netherlands)	www.scopus.com	Largest abstract and citation database of peer-reviewed literature, scientific books, and conference proceedings	>22,000 journals
PubMed (United States)	www.pubmed.gov	Biomedical literature collection from MEDLINE, life science journals, and books	>945,000 articles
SciELO (Brazil)	www.scielo.org	Open access collection of articles, conference proceedings, and books, focusing on central and south America, published by UNESCO	>31,400 journals
Redalyc (Mexico)	www.redalyc.org latindex.unam.mx	Scientific publications from Latin America, Spain, and Portugal	>930 journals; >365,000 articles
MAIK (Russia)	www.maik.ru elibrary.ru	Scientific publications from Russia	
IndianJournal (India)	www. indianjournals. com	Scientific publications from India in agriculture, applied science, library science, management and medical sciences	>355 journals
IRGrid (China)	http://www. irgrid.ac.cn	Collection of Chinese scholarly articles	>1,118,004 articles
BASE (Bielefeld Academic Search) (Germany)	www.base-search. net	Bielefeld University Library open access web resources collection	>65,000,000
DOAJ open access source (United States)	Doaj.org	Article-level search of journals registered in open access database	>1.8 million articles from >6000 journals (open access); >10,000 (all journals)

underlying assumptions about what constitutes publication as well as the use of different databases, counting social science and humanities with the natural science, and rules of inclusion in the indexing services.

The survey of scientific publications suggests that these numbers are under-counting the total world journal publication. Within the BRICs, as many as 15,000 scientific publications are published by national sources, including journals, bulletins, newsletters, and conference proceedings. The extent to which these are academic, refereed journals was not determined by existing surveys. My survey resulted in a count of more than 15,000 national publications in the natural and engineering sciences (excluding social sciences and humanities). Many publications of national character address regional and local issues and problems, such as fish counts in regional areas or water quality report for agriculture. These types of publications will not be of broader interest to a disciplinary community (except as metadata) and therefore would not appear in the abstracting services publishing elite academic research.

An expectation was established based upon the eponymous "Garfield's Law" of concentration. Eugene Garfield (1971) suggested that a basic concentration of journals is the common core of all fields. In his monograph on citation indexing, Garfield used the physical analogy of a comet, whose nucleus represents the core literature and whose tail of debris and gas molecules widen in proportion to the distance from the nucleus—the long tail depicts the additional journals that sometimes publish material relevant to the subject in the core. According to Garfield, his law of concentration postulates that the tail of the literature of one discipline largely consists of the cores of the literatures of other disciplines. This would lead one to expect that a small percentage of all journals could represent highly cited science.

Garfield's Law was posed as the corollary to Bradford's Law of Scattering (1985) which states that a relatively small core of journals will account for as much as 90% of significant literature as determined by citations. Thus, an index that includes a limited number of venues will capture the vast majority of key scientific work in that field. Bradford's Law provides justification for carefully choosing a limited number of publications for indexing. This would suggest that a small number of journals will capture the majority of citations and that Science Citation Index Expanded (SCIE) could contain a small number of highly cited journals, and still capture the vast majority of all citations. It is probably the case that only a small percentage of all published work is read and cited by others.

The literature on the percentage of journals not indexed SCIE is sparse. Charles Morris estimated that 29% of all active, refereed journals listed in Ulrich's periodicals are also indexed in SCIE. To test this, I examined the references made within articles in the Thomson-Reuter's Journal Citation Report (JCR) for 2008, to derive a list of non-source journals. Aggregation of all 2.5 million references within JCR (2008) produced a list of about 300,000 journals not listed in SCIE in addition to the 8207 source journals. A random sample found that about 14% of citations were to publications that are not journals—mainly books and theses. This suggested that the number not indexed in SCIE could number about 250,000 journals. How many journals are actively published? It is still difficult to estimate the number, even with these calculations.

The survey of scientific publications suggests that these numbers are under-counting the total world journal publication. Within the BRICs, as many as 15,000 journals are published, some in national languages and some in English for a broader distribution. Many of the publications in local languages are regionally relevant topics the probably do not have a broader audience. The extent to which these are academic, refereed journals was not determined by existing surveys, but most of them report on local research of some kind. My survey found that a very small percentage of all published works are indexed in sources like Web of Science or Scopus.

A Science-Metrix study (2018) examined the availability of articles by domains of scholarly activity in an attempt to measure the extent of open access (OA) publishing by field. Their research showed that in 2018, health sciences had the highest number of articles available through OA formats (at least 59% of the articles published in 2014 could be read for free in 2016), followed by the natural sciences (55%), applied sciences (47%), economic and social sciences (44%), and arts and humanities (24%). This spread of numbers by field is in part a reflection of the average number of authors on articles: the more authors on an article, the greater the probability that one of them will have funds to pay article-processing charge (for non-free gold OA) and that one author will take the time to archive the article on the public Internet repositories such as arXiv. org. Health and applied sciences are more likely to have coauthored articles than the arts and humanities.

Science-Metrix also found that at least two-thirds of the articles published between 2011 and 2014, and having at least one US author, can be downloaded somewhere on the Internet for free (not include 'pirate' sites like Sci-Hub) as of mid-2016. In the case of Brazil, the proportion reaches 75% of articles available free online. More broadly, the vast majority of the scientifically advanced countries have more than 50% of their articles published from 2010 to 2014, freely available for download in gold and/or green gratis OA.

Confirming Garfield's Law of Concentration

Comparing the number of source journals in SCIE (8207 in 2008) and the number of non-source journals (approximately 250,000 in 2008), an estimate can be derived that SCIE includes about 3% of all scientific and technical journals. Moreover, within the 3%, it is possible to identify 37,067,573 cited references within articles in the source journals; these are divided as follows: 26,809,415 (72.3%) to source journals and 10,258,158 to non-source journal titles. The non-source journal titles also include a large number of misspellings (and name changes), confounding any attempt at a valid count. Nevertheless, the counts appears to confirm Garfield's Law of Concentration in that the source journals garner 80% or more of all citations.

The SCIE data for international collaboration in science, when considered as a network, shows continued to grow in number of nodes and links at a rate even faster than expected. The number of countries whose addresses appear in the network has grown to 201 countries up from 172 countries in 1990.[3] The number of unique documents that have authors from more than one country has doubled. Nearly all nations of the world have become involved in some form of international collaboration. In short, something important is happening to science, and it is happening at the global level.

Table 2.2, originally published in PLoS One, shows the growth of internationally collaborative articles in the Web of Science. The data show that, of all co-authored articles, the percentage that was internationally co-authored rose from 10% to 25% by 2011. Table 2.3 shows the top scientific countries in terms of their spending, output, and the fractional field weighted citations to their work at an aggregated level.

Table 2.3 shows the top scientific countries in terms of their spending, output, and the fractional FWCI to their work at an aggregated level. The countries are listed in alphabetical order. Column 2 shows all publications for this country for the year 2013 as abstracted by Scopus, the Elsevier

---

[3] At the time, the Soviet Union was counted as a single nation.

Table 2.2   Publications in the Web of Science by year, number of journals, and counts of internationally co-authored records

	1990	2000	2005	2011
Relevant records	508,941	623,111	734,750	787,001
Number of journals	3192	3745	3722	3744
Number of authors	1,866,821	3,060,436	3,301,251	4,660,500
Addresses in the file	908,783	1,432,401	1,696,042	2,101,384
Internationally co-authored records	51,601	121,432	171,402	193,216
International addresses	147,411	398,503	618,928	825,664
% internationally co-authored records	10.14	19.49	23.33	24.55

Source: Wagner et al. (2015)

database. Fractional publications listed in column 3 is the share count of all articles that are co-authored by that country; a paper with authors from three countries could count as one-third or 0.33 for the countries included in the address line. Column 4 shows the fractional FWCI, which is a fractional count of citations weighted by the citation levels by field or discipline. Column 5 shows the amount of funds that government provided to R&D in 2011, as reported by the OECD. Column 6 shows the percentage of all publications from that country that is internationally co-authored. Column 7 shows the share of that nation's researcher population that was "mobile" at the international level in 2013—meaning that people went abroad to study and/or work or entered that country to study and/or work.

An analysis of these data showed that research funding is strongly correlated to publication output: the more money spent by national governments, the more articles that are produced. This finding holds true for total government spending as well as per capita spending. For the most part, nations that spend money on research and development can be expected to produce a proportional share of papers. Table 2.3 shows the numbers of articles in 2013 and the spending in 2011, and a quick scan reveals the correlation.

Holding the citation impact index (FWCI, column 4, Table 2.3) as the dependent variable, we did not find a strong relationship between public spending and citations. There is a no significant correlation between spending and citations. More funds spent and more articles produced do not necessarily result in more citations accrued by a country. This is exemplified by China's position of having strong output but fewer citations than

Table 2.3   Numbers of articles published by leading scientific countries, 2013, aggregated citation strength, percentage of all publications that are internationally co-published, and government R&D spending in 2011

Country	All publications in 2013, fractionally counted	International co-publications, fractionally counted	Percent international co-publication	Aggregated field weighted citation strength, fractionally counted (FWCI)	Government spending on R&D in 2011, (GBARD) USD/PPP
Australia	53,706	14,401	27	1.32	4748
Austria	12,808	4483	35	1.15	2921
Belgium	17,636	6459	37	1.34	2880
Brazil	51,107	7193	14	0.66	11,200
Canada	62,178	16,427	26	1.23	7737
China	413,831	37,897	9	0.75	33,268
Czech Republic	14,311	2810	20	0.85	1936
Denmark	13,320	4342	33	1.42	2483
Estonia	1622	433	27	1.06	246
Finland	11,528	3666	32	1.28	2307
France	75,116	22,509	30	1.08	19,984
Germany	106,563	29,314	28	1.21	30,103
Greece	12,120	3078	25	1.01	909
Hungary	6529	1635	25	0.77	666
Ireland	7356	2288	31	1.20	123
Israel	11,995	3015	25	1.12	1480
Italy	68,950	16,944	25	1.27	12,075
Japan	107,008	13,871	13	0.82	34,105
Latvia	1270	195	15	0.72	60
Lithuania	2342	406	17	0.71	100
Luxembourg	861	446	52	1.35	279
Mexico	13,700	2906	21	0.62	5400
Netherlands	33,032	10,843	33	1.45	5951
Norway	10,807	3388	31	1.22	2474
Poland	30,098	3969	13	0.76	2688
Portugal	14,472	4152	29	1.03	2814
Republic of Korea	60,986	8610	14	0.95	17,424
Russian Federation	38,590	5600	15	0.52	18,097

(*continued*)

Table 2.3   (continued)

Country	All publications in 2013, fractionally counted	International co-publications, fractionally counted	Percent international co-publication	Aggregated field weighted citation strength, fractionally counted (FWCI)	Government spending on R&D in 2011, (GBARD) USD/PPP
Singapore	11,402	4111	36	1.68	7600
Slovenia	4068	929	23	0.79	352
Spain	58,079	14,622	25	1.05	10,155
Sweden	20,833	6949	33	1.36	3276
Switzerland	22,441	9047	40	1.48	3611
Turkey	32,277	2896	9	0.66	4581
United Kingdom	107,830	30,213	28	1.35	12,902
United States	449,811	79,209	18	1.40	144,379

Source: Elsevier OECD

other leading scientific powers. Further, this observation suggested that some countries are gaining more benefit by participating in international interactions than others.

In work conducted together with Koen Jonkers of the European Commission, we found that the more open a country is to collaboration and mobility, the higher the impact of their work as measured by weighted citations (Wagner & Jonkers 2017). In other words, more "open" countries are garnering more citations to their work. The fractional FWCI has a significant positive correlation with the percentage of internationalization ($R = 0.53$) and a strong significant negative correlation with the share of immobile researchers ($R = -0.66$; in both cases the p-value <0.005). The other pairwise correlations coefficients were non-significant. The countries that are gaining the most from collaboration are those that are most open to exchange, collaboration, and mobility.

The findings suggest that the nations with researchers who are more willing to engage with the world are also those countries whose work is making the greatest impact as measured by citations. While we recognize that impact is not always the same as quality, it is the same as engagement:

people are paying attention to the work being produced across national boundaries. This finding may help explain recent puzzles about the shifting positions of nations in terms of output and impact. Those nations that are relatively lower than others in terms of openness appear to be lagging in terms of impact. Japan, in particular, has seen output and citation impact remain flat since 2000 (Adams 2010) while also being among the least internationalized of leading nations in terms of both collaboration intensity and inbound scientific mobility flows. The lack of international engagement may be dragging on Japan's performance. Writing in 2010, Adams noted that Japan has a well-established research enterprise and world-class universities: "...[so] it is puzzling to the observer that the average rate of citation to its research articles in the internationally influential journals ... is significantly below...[other nations]." The puzzle may be answered by Japan's lack of "brain circulation" and limited attraction to foreign talent.

Other nations, too, have reduced their global engagement and seen citations drop. Russia, in particular, has fallen in terms of spending, engagement, and citation impact over the 20 years between 1990 and 2010. Italy and Greece also show less tendency to participate at the international level than other European countries. The United States and China also show relatively lower percentage share of participation in global research than other countries, possibly due to the enormous size of their domestic research communities. However, it is notable that neither of these nations is among the most-cited nations.

In contrast, small nations with enhanced global engagement have seen significant jumps in citations. Notable among these—and in addition to the well-known leaders of Switzerland, the Netherlands, Denmark, the United Kingdom, and Sweden—Portugal, Belgium, and Austria stand out as nations that have increased their global reach, with enhanced attention to their research.

An expectation of publishing priorities in journals was established based upon the eponymous "Garfield's Law" of concentration. Eugene Garfield (1979) suggested that a basic concentration of journals is the common core of all fields, suggesting a clustering effect around key, leading journals. This would lead one to expect that a small percentage of all journals could represent much of highly cited science.

The literature on the percentage of journals excluded from SCIE is sparse. Charles Morris estimated that 29% of all active, refereed journals listed in Ulrich's Periodicals are also indexed in SCIE. To test this, I examined the references made within articles in the Thomson-Reuter's JCR for

2008, to derive a list of non-source journals. Aggregation of all 2.5 million references within JCR (2008) produced a list of about 300,000 journals not listed in SCIE in addition to the 8207 source journals. A random sample found that about 14% of citations were to publications that are not journals—mainly books and theses. This suggested that the number not indexed in SCIE could number about 250,000 journals.

Law of Concentration of Journals

Comparing the number of source journals in SCIE (8207 in 2008) and the number of non-source journals (approximately 250,000 in 2008), an estimate can be derived that the Web of Science includes about 3% of all scientific and technical journals. Moreover, within the 3%, it is possible to identify more than 37,000 cited references within articles in the source journals; these are divided as follows: 26,000 (73%) to source journals and 10,258,158 to non-source journal titles. The non-source journal titles also include a large number of misspellings (and name changes). This analysis appears to confirm Garfield's Law of Concentration in that the source journals garner 80% or more of all citations.

The global network shows signs of being an open system in that new entrants can join it; many redundant connections occur within it— ones that enable connections and encourage flow of knowledge, as well as providing opportunities for local search–connecting to others who may have important resources to share in ways that advance research. The network is much less clustered than the World Wide Web, and this is a good sign, especially for developing countries that have plans to use knowledge-based innovation as the means to grow their economies. This is the feature of the global system we explore in the next chapter.

References

Adams, J. (2010). Science Heads East. *New Scientist, 205*(2742), 24–25.
Adams, J. (2013). Collaborations: The 4th Age of Research. *Nature, 487*, 557–560.
Bernal, J. D. (1939). The Social Function of Science. *The Social Function of Science*.
Borgman, C. L., Wallis, J. C., & Enyedy, N. (2007). Little Science Confronts the Data Deluge: Habitat Ecology, Embedded Sensor Networks, and Digital Libraries. *International Journal on Digital Libraries, 7*(1–2), 17–30.

Bradford, S. C. (1985). Sources of Information on Specific Subjects. *Journal of Information Science, 10*(4), 173–180.

Bush, V. (1945). *Science, the Endless Frontier: A Report to the President.* Washington, DC: US Government Printing Office.

Garfield, E. (1971). *Essays of an Information Scientist* (Vol. 1, pp. 222–223), 1962–1973; Current Contents: #17.

Garfield, E. (1979). Current contents: Its impact on scientific communication. *Interdisciplinary Science Reviews, 4*(4), 318–323.

Gokhberg, L., & Nekipelova, E. (2002). International Migration of Scientists and Engineers in Russia. In *International Mobility of the Highly Skilled* (pp. 177–187). Paris: Organisation for Economic Co-operation and Development.

Harnad, S. (1997). The Paper House of Cards (and Why It's Taking so Long to Collapse). *Ariadne, 8*, 6–7.

JCR. (2008). *Journal Citation Reports* (Science Edition). Clarivate Analytics.

Jinha, A. E. (2010). Article 50 Million: An Estimate of the Number of Scholarly Articles in Existence. *Learned Publishing, 23*(3), 258–263. https://doi.org/10.1087/20100308.

Larsen, P., & Von Ins, M. (2010). The Rate of Growth in Scientific Publication and the Decline in Coverage Provided by Science Citation Index. *Scientometrics, 84*(3), 575–603.

Mabe, M. (2003). The Growth and Number of Journals. *Serials, 16*(2), 191–198.

National Science Board. (2016). *Science & Engineering Indicators.* US Government Printing Office.

Organization for Economic Cooperation and Development. (2018). *Main Science & Technology Indicators.* https://data.oecd.org. Accessed July 2018.

Price, D. D. S. (1961). *Science Since Babylon.* New Haven: Yale University Press.

Toole, B. A. (1992). *Ada, the Enchantress of Numbers: A Selection of Letters from Lord Byron's Daughter and Her Description of the First Computer.* Sausalito: Strawberry Press.

UNESCO Science Report 2010. (2010). Geneva. Retrieved from http://www.unesco.org/new/en/natural-sciences/science-technology/prospective-studies/unesco-science-report/unesco-science-report-2010/

Wagner, C. S., & Jonkers, K. (2017). Open Countries Have Strong Science. *Nature News, 550*(7674), 32.

Wagner, C. S., Park, H. W., & Leydesdorff, L. (2015). The Continuing Growth of Global Cooperation Networks in Research: A Conundrum for National Governments. *PLoS One, 10*(7), e0131816.

CHAPTER 3

Levels and Patterns of Communication in the Global Network

Science is conducted to address complex problems in step-by-step processes. Each step, carefully recorded, is added to a body of knowledge.
Even disappointments and "dead-ends" provide useful information. The
results are known because they are communicated before, during, or after
the work. The communication process, carefully crafted into text, at first
for knowledgeable peers, and then for a broader scientific community, is
the essence of science, and represents the process described by William
Gass (2006): "The true alchemists do not change lead into gold; they
change the world into words." In a model of the scientific world, we
would consider the words as the base layer (the chemistry of its social life)
upon which all else is built. It is why a common language has always been
so important to science.[1] This chapter describes the communications
dynamics within the global network of science as drawn from the journal
sets indexed in the Web of Science.

Consider the case of the search for the CRISPR-cas9 described by
Jennifer Doudna and Samuel Sternberg in "A Crack in Creation: The New
Power to Control Evolution" (2017). The authors describe a process of
inquiries, dead-ends, cooperation, collaboration, and the publication of
dozens of papers over a number of years. Doudna described the process as
"a great story of how curiosity-driven research aimed in one direction ended

[1] The common language of science historically has been Greek, Latin, German, and now
English. Perhaps Mandarin Chinese is next.

© The Author(s) 2018
C. S. Wagner, *The Collaborative Era in Science*, Palgrave Advances
in the Economics of Innovation and Technology,
https://doi.org/10.1007/978-3-319-94986-4_3

up uncovering something that could be employed in a completely different way. I think that the way that bacteria can program proteins to cut viral DNA and protect themselves from viral infection was the original work that we were doing. This was a project—an international collaboration—with Emmanuelle Charpentier [in France] and her laboratory. And that uncovered the mechanism that we realized could be employed in a very different way, namely for gene editing" (Radio interview, 2017). This story exemplifies adaptive behaviors in action.

During the processes of executing the scientific method—developing a hypothesis about how CRISPR changes the genetic code, in crafting experiments with viruses, attracting collaborators such as Dr. Charpentier, and reporting the results—communication processes were clearly an essential component to Dr. Doudna's work. The communication process is so basic that people rarely stop to consider its importance. Certainly, researchers are exquisitely careful in the words they use to describe their findings, since common understanding among peers is clearly needed. Sometimes, new terms are needed, and these are invented and diffused in the literature. Scientific knowledge depends upon a multi-leveled system of communications from informal exchanges to highly formal publications. More importantly, for governing the global network, the system of communications can be studied to reveal a social structure. This chapter discusses the complex communications taking place in the global networks, and what we can learn by studying them.

Taking a long look back, historically, the communications system of science did not begin as a complex system, but as a linear, functional one of communicating on paper, of sharing notes, letters, and pamphlets. The original experimentalists of the seventeenth century, the invisible college, wrote letters to one another with reports and updates on their experiments. The letters were eventually collected into abstracts; fuller treatments were placed into journals. In 1731, the Royal Society of Edinburgh introduced peer review as a requirement prior to the publication of medical papers. By the mid-eighteenth century, journals had been established as the primary method of communicating science.

Why do knowledge systems emerge, and why do they evolve into more complex systems? Elinor Ostrom and Charlotte Hess (2005) addressed this question within the context of knowledge as a commons—a shared resource—that flows from one person to another, creating public value. The idea of the knowledge commons fits well for scientific knowledge in the current era. Ostrom and Hess tag articles as "artifacts" that contain the expressions of ideas. They define ideas (also, for our practical use) as coherent thoughts, mental images, creative visions, and innovative information: "The most notable characteristic of an idea is that it is a pure public good and, therefore, nonrivalrous. One person's use of it does not subtract from another's."

Ostrom and Hess pointed out that open access in the digital knowledge commons has greatly shifted the idea of "location," moving science from the paper library (where information could be rivalrous) to electronic, distributed information (where knowledge is non-rivalrous). This shift alone, without the added international dimension and knowledge diffusion, serves to change codified information from a rivalrous toward a non-rivalrous good within a knowledge commons. Moreover, the nature of the "commons" shifts knowledge from a scarce resource to an abundant one. The actors within the network of science communications have responded to the changes in their environment; as information has become abundant, researchers who once maintained a tight hold over information shifting to open sharing and broader access, even to pre-publication data. Like the shift in the seventeenth century from the mysterious and secret world of magic and alchemy to the reproducible codified article, the shift now is from scarce, rivalrous information to broadly shared information across a globally accessible knowledge base.

In the process of moving from linear growth to exponential growth in knowledge production, and from scarce, rivalrous books, journals, and paper copies available in a few elite libraries to abundant, web-based open access, the knowledge system has taken on the features of a complex adaptive system (CAS). As this has happened, the researchers accessing, creating, and sharing knowledge have adapted to the changing environment, moving from individual research projects to teams and collaborations. The teams and collaborations become parts of networks, resulting in the emergence of the global network, as we will discuss.

THE COMPLEX ADAPTIVE SYSTEM OF GLOBAL SCIENCE

The term "complex adaptive system" was coined at the Santa Fe Institute in New Mexico by John H. Holland. In *Hidden Order*, Holland (1995) describes how adaptation by interacting agents builds complexity, and how complex organization, or order, can emerge from seemingly simple agents operating with just a few rules. Holland's idea anticipated by Herbert Simon (1962), who observed that complex natural systems naturally develop into hierarchies. In describing complex systems, Simon found them to be composed of interrelated subsystems that layer from simpler to more complex, such as from an atomic level, through a chemical level up to include biological tissues, organs, and organisms. Each layer is subordinated to the one above it. Ordering within layers creates a necessary condition for emergence of natural phenomena at the next level.

The same layering and emergence can be found in social systems, but with specific differences from natural systems. Social systems also often organize as hierarchies. Hierarchies are effective organizations for production of a good, or for command and control organizations that need clear direction, especially when combined with followers who are expected to take direction. The twentieth-century corporation is an example of effective and efficient hierarchies.

When exploration and research is a goal, social organizations take a less fixed form than hierarchy. Researchers need freedom to search for new ideas, tools, and mechanisms of research, as well as to align themselves with the depth of knowledge needed to explore a topic. A good deal of expertise and a strong base of conventional, or accepted, knowledge is needed. When the task is problem-solving, the ability to explore new ideas is critical to scientific advancement. Figure 3.1 shows three different types of human organization: hierarchy, heterarchy, and network. We are very familiar with hierarchical social structures, so let's turn to heterarchy and network, which are important to understanding global science. In a hierarchy, as Simon described, social systems contain subsystems that may have no relation of subordination; in other words, the local components do not necessarily "answer to" or "take direction from" the other parts of the system. Within social systems, layers exist, but communication can occur both within and across layers. Basic rules of interaction determine how communications occur. The communications can influence how layers organize. The multiple layers and intercommunication create conditions for what Kontopoulos (2006) called a heterarchy. The heterarchy model expands upon Simon's view to better explain social organization, again, shown in Fig. 3.1.

In a social heterarchy, we find multiple layers. Each subsystem is related to the one below and above it, similar to Simon's construct, but with an ability to adapt to new information and resources. This layered organization creates a complex several-to-several or many-to-many relation across levels, shown on left in Fig. 3.1. Kontopoulos (2006), who interpreted Simon for social systems studies, observed that, in some systems, the subsystems are not subordinate to others; in other words, they interact and influence one another. Thus, the subsystems can be examined as nested

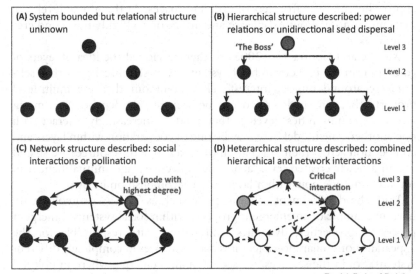

Fig. 3.1   From network to heterarchy in knowledge-creating groups

levels, with feedback loops among them. Partial inclusion between levels (labs, institutions, regions, nations, disciplines, and so on) introduces autonomy at each level as well as a "crucial form of tangledness" (p. 54). Kontopoulos (2006) termed these forms of nested and interactive systems heterarchies (tangled composite structures)—forms that include levels with partial determination from below, partial determination from above, and semiautonomy on each level.

For our purposes, Kontopoulos' heterarchy model improves upon Simon's hierarchy of nested layers in that, for science at least, we see no single governing level or expectation of subordination at the various levels. The various levels exert an influence on each other; at each level, they are characterized as having multiple access points, multiple linkages, and multiple determinations. The global science system begins with the simple rules embedded in the practitioner that includes deepening conventional knowledge, searching for resources, looking for new ideas and recombinations of knowledge, and seeking attention for one's contributions. At the disciplinary level, groups seek consensus around problem definitions as well as agreement about a core literature, acceptable approaches, and the frontiers of problem spaces. At the global level, the dynamic is more about excellence and attention to highly novel and creative work.

Monge and Contractor (2003) further developed the idea of layers of communications in heterarchical systems by describing layers that self-organize around mutual interest. They point out that the multi-level property that Simon described can be seen in interdependence among layers. As communities develop, they tend to increasingly interact with one another. The model here may be one of cooperation within layers and competition across them, in that mutual exchange of resources are shared within a newly created level, achieving collectivity and mutual benefit in ways that no one unit-within-layer could accomplish on its own.

With the development of a discipline such as biology or geology, we have an example of an emergent property within social systems. Emergent processes, properties, dynamics, and patterns are responsible for the appearance of similar global features and structures in completely different systems. So, in the sciences, we see disciplines emerge, where in political systems we see factions, or in social life we see cliques or niches. Emergence and adaptation are characteristic features of many complex, social, and self-organizing systems. In the sciences, the emergence of a discipline (or new subdiscipline) can be seen as the dynamic process of reorganization based upon new information or resources. Adding a new telescope to view the heavens brings new information that may encourage group adaptation and reorganization to take advantage of new data and to group into a subdiscipline, perhaps one studying exoplanets.

A system has high complexity or is very complex if it can be represented efficiently by different models at different scales. This is part of the challenge to governance posed by the global network. The overall complexity is determined by the number of levels on which structures can be found. In global science, we can identify at least three emergent levels (practitioners organizing into teams, disciplines to which individuals associate, and networks of interactions) without considering the institutional levels, which we will discuss later. Emergent levels within a multilayered system can have scale-free properties or fractal or self-similar organization, and we will consider this phenomenon, as well.

In a complex system, there is local interaction. But, as we see, viewing the local actor does not reveal the nature of the system. As John Padgett and Walter Powell (2012) pointed out, "in the short run, actors create relations; in the long run, relations create actors." Examining the actions of an individual practitioner does not reveal the dynamics of the whole network. However, we see that the global network now provides access to these flow resources in the knowledge commons, while also setting the boundaries of interaction. As such, the global network has recognizable features in common with other CASs. The most notable, as we have seen, is its emergence in the absence of a governing institution, a feature shared with phenomena like language, weather, and even life itself, according to biologist Stuart Kauffman (1996). *Emergence* is the term for the process by which complex systems arise spontaneously out of a multiplicity of relatively simple interactions. This concept helps us describe and understand the global network, since that network is not constructed or institutionalized. It is entirely emergent based upon the interactions of actors, ideas, resources, and goals bubbling up from lower levels.

The mix of ideas embedded seen within published artifacts spontaneously crystalize into systems we call theories, paradigms, or disciplines. That is the CAS formed out of the interaction of its facilities, artifacts, and ideas in the knowledge commons, forced into existence by the "organisms" (practitioners), their individual motivating actions, and their environment, organizing into groups rather than in response to any specific set of directions.

We understand the global network of scientific collaborators to be, not the result of actors or entities consciously building it, but as emerging from the search for competence, resources, new ideas, and attention. Each scientist simply goes about his or her own business, focused on advancing their career, asking interesting questions, writing, teaching, and conferring. In seeking attention, and in choosing to exchange goods, they participate in social dynamic that operates as a vortex, pushing upward in attention the most notable members.

This basic fact, that each member of the network is seeking to maximize his or her own welfare, that the scientist pursues his or her own self-interest without much regard for the whole, would seem to be a recipe for disorder and chaos. Just as the "invisible hand" of the economy conditions order, the global system of science emerges, even against the countervailing forces of national interests, different languages, time zones, and social norms that constitute the seemingly harsh and unkind environment. Indeed, in a beautiful paradox, the interactions in pursuit of self-interest may simultaneously advance not only individual goals but also the welfare of the system and the state.

We tend to understand CASs intuitively because we observe and experience them. We are intimately familiar with the complex systems we live within, even without giving it much thought—the human body is a complex ecosystem, with consciousness as the emergent feature, even if we do not stop to think about ourselves in that way. We can study systems at many levels, but isolating any one feature cannot provide insight into the whole of an entity. In other words, the system cannot be reduced to its component parts and still be understood as a system. Examining an oak tree does not reveal the functioning of the forest; examining the brain does not reveal consciousness. And usually it is the whole of the ecosystem that provides the benefits that are of interest, or at least in this case it is: we want to understand the ecosystem of the global network of science.

The elements of complex systems that make them interesting (their unexpected features, their evolution, their ability to adapt) also make them difficult to model and measure. Natural ecosystems are dynamic, with all parts interacting and influencing each other, each in their own ways on their own time scales. Subsystems also change in relationship to each other, as we see between animals and climate change. This also applies to the global science network. It consists of quantities of matter, energy, and information in ways whose contributions, interactions, and outcomes are not completely understood.

The global science network is similarly difficult to measure and understand. It is nearly impossible to sort out the flows of knowledge, people, and money. Yet, it has developed enough stability *as a system* to warrant study as a network. Not only is it more feasible to study the efficiency of the system than to try to unravel the threads of each part of the flow resource, happily, it is also more informative to study the system in this way. Just as with other CASs, studying just the individual components (the trees) will not tell us what makes the global network operate (the forest). We need to study the global communications as a system.

Systems studies are not new in themselves—science has studied systems for a long time. We have come to see systems as having multiple layers that become integrated, often beginning with simple components (if we think of biology, we know that biological systems depend on a chemical level, for example) and from simpler levels, developing into higher levels of more complex organization. Each level of organization has a greater complexity and a higher degree of freedom than the simpler level below it, but the upper levels also use more energy and have more degrees of freedom. Think of the human body as having a chemical base upon which biological levels are built, from simple cells, complex collection of cells (tissues), systems such as the nervous system (organs, like the brain), and finally the emergence of consciousness—perhaps the most amazing (and dimly understood) emergent property on earth.

Advances in computer modeling and mathematics have revolutionized systems studies. Many more fields of science are able to model CASs such as biological ecosystems, solar systems, and weather systems. From these models we have begun to derive underlying rules and structural dynamics by which complex systems operate and evolve. These advances give some hope for better understanding of systems. The same possibility of improved understanding can be extended to the complex system of science as well. To explain how that might work, I will use the next section to discuss ways in which the global science network shares conditions with other CASs.

THE SYSTEM DYNAMICS CREATING THE GLOBAL NETWORK OF SCIENCE

Like other complex systems, the study of the global network of science cannot be made solely by reference to its component parts. Just like the components of a coastal ecosystem—which includes aquatic life, sand, grass, water, and microbes—tell us little about the health of the coast, the individual components reveal next-to-nothing about the system. Raindrops do not explain a hurricane, neither does examining the activities of the individual scientist working in the lab or in the field tell us much about the whole of science. The scientist can work for a time unaware of the larger knowledge system to which she contributes. He or she may have personal motives for choosing work that has little to do with the system of science as a whole. Moreover, he or she is not working in the lab in order to win a Nobel Prize, or because Einstein wrote about relativity. The individual scientist has many different motivations for taking action, most of which have no relationship to the rewards given to scientists.

The global network is not hierarchical even if some scientific activities are captive within institutional hierarchies. One layer of organization within it does not cause conditions for other layers. No overarching group creates conditions or directions for the global system. As such, the global network has levels of activity that will be presented as layers that are nested within each other in terms of complexity rather than power. Complex systems—heterarchies such as science—fall in between other systems that range between simple input–output systems to the seemingly unfathomable universe. A distinguishing feature of heterarchies is that they are autonomous (in the sense of independent, self-governing) systems that operate by internally determined rules. The primary distinction of autonomous systems is that they cannot be characterized by inputs (stimuli, raw materials, money, or energy) but must be understood by examining the interrelationship of components and the larger structure, which they build through the communications process.

Here, we turn to a more controversial aspect of network and communications theory: that the macro-behavior of a network is disconnected from the behavior of the agents. In other words, the global network has its own rules of operation that do not rely on the intentions of the participants. Rather than relying on the participants for form and function, the formation and persistence of the network is the functional, organized complex system operating according to its own dynamics. The actors may have created the relations, but once a dynamic system is in place and stabilized, the relations create the actors.

The network structure provides resources and opportunities to participants; these attract practitioners to seek participation in the network. The network structure also enforces constraints. To enter the social network means that entrants must be attractive in themselves and offer to reciprocate goods of some kind to join the network. (Reciprocity can be asymmetrical, and as basic as an intriguing question or local natural phenomenon.) Reciprocity as a core feature of network structure may be the reason that capacity building (rather than information technology) is more influential on the development of the global network, as we will discuss later.

The collaborative structure of the network, and the rules by which one joins and participates in the network, also solves a long-term problem associated with collaboration—that of the "free rider" in teams. Evolutionary biologists assert that cooperation and collaboration are the social functions that created the need for the bigger human brain that evolved in *Homo sapiens*. In an article on collective action and the collaborative brain, Gaviet reveals modeled results that show that competition between groups and collaboration within groups are factors favoring high intelligence. Within a group, the entire payoff is dependent upon what the others do. It is both more costly to maintain a long-distance relationship and also more rewarding because if one person defects from the group, it is clear right away, whereas in proximity, it may not be clear that a member is no longer cooperating. At a distance, cooperation among the group is self-monitored and free riders are quickly purged. This means that the long-distance cooperations are less likely to drag along free riders or uncooperative partners, and therefore have the potential to be more likely to be highly productive.

THE INTERRELATIONSHIP OF SCIENTISTS AND THE GLOBAL STRUCTURE OF SCIENCE

Disciplines, fields, and subfields are one kind of layering in scientific ecosystems. Another operating feature is geography and location. Researchers operate within a dispersed community. We live in a world in which the number of places conducting good science is spreading rapidly to new places. (Some of these places are growing their own ecosystems of actors.) Laboratories, research centers, industrial labs, government labs, and universities are proliferating in number, and they are located in many parts of the world, some regions of which (such as in Asia and Africa) are new to science. Each of these locations has researchers who are responding to the abundance of knowledge available—that knowledge that was difficult to access in the past, but which has become more readily accessible with digitization. They are also responding to the opportunities to join networks of researchers. Each geographic group represents layers of organization that span space and include people who self-identify into groups of dispersed communities created by people who wish to be identified as part of a discipline, field, or subfield. They learn the norms and language of the subdiscipline and seek to be recognized as part of the group.

Similarly, expertise and input into research is drawn from many more places in society than in the past. The greater numbers of citizen scientists being reported in the news is one example. Just as consumers provide feedback to producers to help improve products, scientists may take feedback from citizens back into the lab to try out new solutions. As has been pointed out by many others, such as Eric Von Hippel and Esther Dyson, society is moving into a position where people are able to communicate wishes, desires, and fears to science with the hope that science will respond, hopefully with improvements. This perhaps confirms a Mode 2 theory (see Chap. 1) of operation where we expect to find a feedback loop between science and society that is beneficial to both sides, and which results in a science ecosystem that often seeks to address social needs. In the process, the additional layer of the social component adds a new layer to the complexity of communication and experimentation in science. Science is increasingly engaging with the consumers of science. For example—whale and bird sightings are collected by citizens and reported to a centralized portal for comparison. Citizen science and crowdsourcing of scientific research demonstrate that the consumers of science are engaging in it directly.

Very often, the behavior of the individual is motivated by the rewards offered in their environment. Ostrom and Hess identify the rewards as a key factor for understanding the knowledge commons. They cite high visibility and potential impact as incentives for participating in the knowledge commons. We know that higher citation counts for published articles can be a factor in other benefits such as funding and promotion. As more and more elites participate in the global network, it becomes more attractive to others to join, too. This becomes a virtuous cycle of attraction and participation. As a rule, the working environment provides feedback to individuals with rewards for a decision or a direction taken; or the feedback may be negative and cause people to change direction or to seek resources and rewards elsewhere. This provides some of the social "steering" of science—a feature that comes in for criticism in some quarters which provides social information to the practice of science.

The initial organization of the network—the stage where actors create relations—is based upon the basic rules and characteristics embedded in the agent. By that I mean, as scientists operate within their environment, they follow their own local incentives for creating patterns of behavior. For example, some behavior is competitive as people seek access to scarce resources. Resources may be allocated to participants based upon their characteristics (aggressiveness, creativity), location (prestigious institution), and age or experience (reputation). Funds and other resources (lab space, graduate students) are often allocated based upon personal characteristics, but the allocation can also influence the rate at which the resource is renewed.

Interactive behavior is also cooperative. In scientific research, practitioners often cooperate. Cooperation can improve the ability to gain and use resources, especially when a team approach or collaboration is an effective way to conduct research. Cooperation includes many different kinds of sharing, which we will discuss in more detail in this chapter, but it can include collaboration in research, teaming with people from different fields, sharing equipment or samples, training one another's students, and many more actions that promote both the individual and the group.

COMMUNICATIVE BEHAVIOR

In complex systems such as the global network, communication becomes the flow of resources that enable interactive behavior. This is pretty obvious—communications are essential whether they are physical interactions (swapping of goods), sharing ideas, or providing feedback. All features of the complex system require some form of communication. The flow resource of communication is akin to blood and connective tissue in the human body, or to the food web within the coastal ecosystem. Whether a scientific practitioner wishes to share discovery, doubt, agreement, criticism, or puzzlement, communication is the mode by which these expressions become part of science. Without putting too fine a philosophical point on it, until a scientific idea is communicated, it cannot be considered part of "science." It may be knowledge (in the heads of people), but it cannot be science: in fact, for science, it is not just the communication of a datum, but the more formal codification of the scientific process —literally put into code like writing or equations in scientific venues—that is measurable, transmittable, and actionable, and thus enables what we would call scientific knowledge.

For the most part, communication in science occurs within groups of researchers who generally share a common jargon developed through years of study. As a rule, few people share the common language of a scientific subfield, say, one such as proteomics where knowing a range of cellular proteins like cysteine, trypsin, and subtilisin are everyday terms within the subdiscipline, but are beyond the usage of most people. Technical sophisticated language is a way of creating a boundary around a group, and allows people a way to identify with a group. It acts (for the most part) passively to exclude some who are not trained and initiated into the subfield. As a shared verbal communication, the specialized language can be broader and more flexible than written language, which is often more formal and highly stylized. The written language makes the information eligible for publication in scientific venues such as conference proceedings, journals, archives, and repositories (like arXiv.org). Mastery of specialized language of a subfield of science—or the lack of mastery—has much to do with determining someone's position within their scientific ecosystem.

Once scientific communication has been codified into text, it becomes part of the communications process by which scientists "signal" one another about the nature of their research. Published works constitute a public record of scientific activity. Scientific publication operates as mostly an open system in that participation is not limited to a set number of people or pre-established limits on publications. Anyone who is willing to learn the language of the subfield may participate by submitting an article to a journal for publication or a conference for consideration. This is easy to say, but in practice it is difficult to achieve. Simply learning the language of science takes years, and then writing the research into an article for publication in an elite journal is even more challenging. A small percentage of scientists even have their articles published in elite journals. These articles are usually submitted by editors to peers for review and acceptance. Once a person publishes or formally transmits work, that material joins a traceable network of communications, a body of knowledge.

Scientific publications have certain norms that are (for the most part) highly consistent. The scientific article is a norm for disseminating knowledge. Articles cite previous research (contained in other articles, proceedings, or books) that has influenced the work through footnotes, endnotes, and references. Sometimes, scientific publications will also contain references for further reading, also citing earlier work. Many scientific publications contain acknowledgments to funding agencies and to people who advised or assisted in the writing or reviewing of the article. A publication will generally have a title that distinguishes it and its content from other publications; and most have an abstract, through which they summarize membership and participation in a particular topic or area of study. Under the influence of search engines, keywords are also provided to indicate broad membership in specific fields.

Scientific communications have evolved into a distinctive genre, with regularities that are examined by communications scholars. Theories exist about how scientists choose title words, keywords, and abstract words. One specialist even proffered a theory of acknowledgment.[2] Similarly, communications theorists have theories about citation and referencing behavior and why scientists cite others in the way they do. For our purposes, these communications theories are less important than the linkages that they represent to help us gain visibility into the complex network of science.

Some creative researchers, for example, Kevin Boyack and Katy Borner, are "mapping" science to reveal the underlying dynamics of the commu-

---

[2] Lee Giles from Penn State has written on how and why scientists acknowledge one another in scientific articles.

nications between scientific participants. The maps take a number of forms, but the most common one is a network. At the simplest level, mapping of science can be done at the level of the researcher by placing him or her into a network of collaborators, with each node in the network representing a person, and lines between them (links) showing a co-authorship or citation relationship.

Mapping can also be done at the knowledge level (such as drawing connections between two articles that commonly cite another article as having a common intellectual predecessor), and at the disciplinary level by showing links among fields of science to one another (this is done with cross-references between journals). Each of these networks can be examined as an ecosystem in that the network will contain smaller subsystems which exchange information and resources.

In this beautiful map of science shown in Fig. 3.2, the authors collected all the citation information from all the scientific journals listed in Web of Science (a database of elite scientific journals) maintained by Clarivate Analytics. This database abstracts articles, notes, and letters from elite journals. The links between the journals based upon references represent the aggregation of the counts of any time one article in one journal references another journal. The network shows the references among them as a knowledge network. The colors of the map indicate the fields of science, and coherence is revealed in that biology melds into chemistry, which blends into physics and so on. The map reveals that there are "core fields" of science such as biology, chemistry, and physics, but between these fields it is possible to see many connections that cross from one field to another, spanning boundaries and representing knowledge connections. Each field has a core of interconnecting references, but at the edges, the field eventually meets its neighbor in a great network of global and universal knowledge. In fact, the map shows that all of science is connected from one end to the other in a structural communications network.

The colors in the map correlate to the fields of science that the National Science Foundation (United States) uses to report on scientific activities. Green represents biology, which links closely through citation behavior to the earth sciences on one side, and to biotechnology and infectious diseases on the other. The links derive from all citations in all the articles as they reference their own field (and subfields) and other fields, as well. In this case, the map is an aggregation of all the citations across all the articles in the Web of Science in 2008.

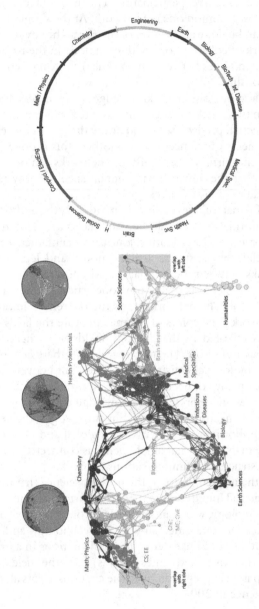

Fig. 3.2 The map of science based upon co-citation relations. (Source: Map of Science)

RESEARCH AS ENERGY FLOW

Continuing the ecosystem analogy, research is a large part of the "energy" or food for the system that in turn creates communications functions within the ecosystem of science. For example, research is the activity represented in the map of science above. In large part, it is the knowledge-creating engine of science—although I should emphasize research is only a part of science, as the latter is a vast body of accepted knowledge (captured in textbooks, norms, and practices). Research is the process of asking questions about the "unknown." It addresses questions it can plausibly answer using the scientific method in a reproducible way. Its proper practice adds to the stock of scientific knowledge—even when findings are negative.

These days, to hold its place in a scientific ecosystem, researchers are often held publicly accountable in several ways and for several reasons. The first level of accountability concerns funding: between one-third and one-half or more of basic research funding comes from taxpayer coffers through government agencies. The second level of accountability concerns safety: science should be conducted safely and the outputs of science should improve or enhance safety. The third level is a question of public welfare: has science contributed (overall) to net social improvement. In short, for modern scientific research, public accountability must be considered as part of the ecosystem as well. It represents important restrictions on the "flow of energy" (research activity). For the researcher, this means there are two costs—costs of research and costs of communication.

Just as conserving energy is important in the natural world, in scientific ecosystems, the individual agents (scientists, labs) attempt to keep the "costs" of communication low in order to pursue their goals. They seek out the most relevant and convenient resources, including knowledge repositories and/or personal collaborations primarily to accomplish these goals. Efficiency of communications is essential to knowledge creation. For example, it is part of the reason for creating a specialized language (jargon) so often associated with scientific fields and subfields. For the scientists, a customized language is much more efficient for exchanging complex information than is using common language.

As I noted earlier, incentives such as reputation and reward play key roles for researchers, laboratories, and scientific institutions. In studying the OA software movement, for example, Eric Raymond found that self-interest was the most common motivation for sharing software—more so than the incentive to create knowledge linkages through exchanges through the "gift economy"—a concept that is sometimes tied to the "OA" movement. Just as in the "invisible hand" of the economy and the invisible college of researchers, benefit at the group level comes from each agent pursuing self-interests. The group benefits from maximum choice available to researchers (actors). In "open source" software, researchers wanted to use each other's software to address the immediate problems they were trying to solve, with the larger goal of enhancing their reputation in their fields. Link between problem-solving and reputation-seeking has been shown to motivate many other kinds of scientific exchanges, as we will discuss.

There is also conservation of energy in the search for new ideas, so people tend to look for nearby practitioners for answers before venturing too far afield. Research shows that people are much more likely to establish connections to those who are physically close, because they come into contact with one another and/or because of the ease of sharing resources. Physical proximity can facilitate the exchange of what Nonaka has called "tacit knowledge"—physical knowledge and experience that is difficult to write down. Sometimes it is easier to demonstrate action—parts of communication that cannot be codified. Highly complex, experiential knowledge (physical knowledge) does not transfer easily, no matter what the proximity or distance, and often, feedback and repetition are needed for people to be able to share a common outlook or way of knowing something. This is the value of apprenticeship and education.

Co-evolution and Change in the Complex System

Charles Darwin's theory of natural selection, explained in his 1859 book, *The Origin of Species*, has been misinterpreted over the years as asserting that evolution is a process of "chance and accident." For many years (in the historical space between the theory of natural selection and the discovery of DNA), it was difficult to counter the "chance and accident" label. Charles Darwin himself noted the gap: "When the views advanced by me in this volume, and by Mr. Wallace or when analogous views on the origin of species are generally admitted, we can dimly foresee that there will be a considerable revolution in natural history." Given their scientific equipment and the pool of scientific knowledge available at the time, many questions about how the natural world changes over time could not be answered by nineteenth-century naturalists, and they themselves acknowledged this fact.

While many of these questions persist, findings in genetics have added richness to Darwinian natural selection, and provided, at a minimum, sophistication to the fundamental understandings of evolutionary change.

More recently, findings in systems dynamics also have helped respond to "chance and accident" accusations by demonstrating that order can emerge from within seemingly chaotic conditions, as it does in science as a whole. The hidden dynamics that give rise to variation and speciation—the dynamics whose processes remained mysterious to Darwin—are better elucidated now. With the advent of large-scale computing, new models are helping provide insight into the basic rules underlying emergent order in complex systems. We have begun to identify the underlying rules of ecosystem function—layered organization, distributed power, adaptation to local condition, search for resources through competition and cooperation, and the importance of communications and interactions as providing structure. In addition, we know more about how systems change over time.

Indeed, the changes and shifts—the evolution—in the system are some of the most interesting parts of studying science and have been the object of many different theories. In the way we're looking at science here, the scientific ecosystems' structure, the "species composition" (in science, the disciplinary fields), and function change over time—they evolve. Sometimes they evolve with regularity, and sometimes as "catastrophes" or phase shifts; Thomas Kuhn called them "paradigm shifts." In natural ecosystems, the changes can appear as cyclical phenomena or as sudden shifts between apparently alternative "stable" states—such as from normal weather patterns to a hurricane. Science also features shifts between states, which is one of the things we will explore in more detail. For one thing, it may be part of the shift from scarcity to abundance in the science system that has brought us to the point of feeling truly overwhelmed by scientific knowledge. In the natural world, environmental fluctuations can cause longer-term change, such as shifts in ocean currents and their influence on growth of flora. In science, shifts can result in long-term emphases of one field over another, such as we have seen with shifts from physics and chemistry toward the biological sciences in the West, as findings in a field such as the synthetic biology produces a string of amazing findings.

The patterns that define scientific communications act as selection pressures on science as new ideas emerge. Certainly, social expectations, problems, and challenges play a role in the choices of projects that receive funding. Other inputs include advances made that reveal anomalies: this has been the case with the discovery of, for example, graphene. This previously ignored relative of graphite was recently resurrected with findings about the electrical properties of the materials. The insights gained attracted a great deal of funding not only for graphene, but with a positive spillover effect on funding for other basic materials research.

As arguably the greatest instrument of adaptation, communications create both structure and interaction across the landscape of the system of science. For example, scientific journals can be said to provide structure in that they serve to identify some knowledge as representing knowledge within a field. The more elite the journal, the more likely it is to publish deeply disciplinary work. Journal editors and reviewers serve as gatekeepers by choosing which articles to publish; they therefore serve to communicate and retain knowledge related to a particular discipline. The journals serve a structuring role in science by defining topics that are within a discipline. It is important to keep in mind that different agents in the same system, including agents at different levels (layers) of organization, may follow different rules. The same agents may follow different rules at different times, and multiple rules may be in operation at any one time, both within a population of agents and across different populations. These differences will become more important to the discussion as we unfold this example of the system of science.

Scientific conferences, which are common and frequent parts of scientific communications, create opportunities for people to communicate research and findings to others with similar interests—as noted at the beginning of the chapter, they are the events seeding most collaborations. Jennifer Doudna describes how an international conference was the place where she met Emmanuelle Charpentier. Doudna was familiar with Charpentier's work, but it was when they met at a conference that they forged a common vision of an international collaboration. Face-to-face communications become part of the connective tissue that ties together the layers operating at different levels of formality, with some communications being highly ephemeral and informal—such as Doudna and Charpentier's conversation—and others being highly formal and persistent, such as a book or textbook. The communications processes represented by this spectrum from informal to formal pass along and reinforce the internal rules of the organization by using highly specialized language of a subdiscipline such as the search for CRISPR-cas9 that only a small number of highly trained and initiated people can understand. A book such as that including Doudna and Sternberg's description of year of research, and its broader social implications, has a reach in scale for a broader readership and scope: a longer "shelf life."

A robust and resilient system will have an environment where both competition and cooperation coexist, or, to express this idea in the language of biology, the ecosystem supports variations and various species. Some species are highly compatible and even symbiotic with one another, while others are competitive, even predatory. The more robust the ecosystem, the more it can tolerate diversity. Consider the difference between the tundra–an environment that can only sustain a few species, and a rain-forest where even a genus of butterflies can include hundreds of variations. Within modern science, research practitioners also operate within invisible colleges—specialties or disciplines—that are emergent, self-organized, and operating oftentimes outside of institutions—creating a network of inter-connections which determine (often unwritten) standards and norms for participation as well as a special language that only the "initiated" can speak, as shown in the figure of Nobel Prize winners.

In science, the delineations among these invisible colleges are fuzzy. Where does biology end and biochemistry begin? The map of science (above) may give some indication but not precision. Many times, the invis-ible colleges have open, porous borders across which many researchers "travel" in their search for answers to difficult questions. Even scientists can have a hard time saying what their field (subfield or niche) is—they are more likely to tell you what they trained to be, not what they do now. This may be an indicator of adaptability of the ecosystem and its ability to change.

Communities within a natural ecosystem are interdependent and inter-connected as networks, depending on environmental variables. Relationships within biomes can include predator–prey (sharks and small fry), symbioses, mutualism, and parasitism. Transferring the analogy to science, the scientist must be able to adapt to local conditions related to his or her research—and this will mean either competing or cooperating, depending upon the circum-stances required at the time. Interdependence presents multiple challenges around the use of and access to resources. In a natural ecosystem, most of these relationships do not involve conscious choice on the part of the species. However, we assume that human participation in a scientific ecosystem has more choice as a component of participation. This is a question worth explor-ing in more detail, since even scientists need to be "fed" in a food-web equivalent that includes financial, material, and knowledge resources that are beyond the control of the individual.

Indeed, in an ecosystem analysis, the food-web relationships are essen-tial for understanding possible states of the system. The interaction between populations and their environment—particularly as it may lead to a hierarchy of sufficiency or abundance—is a critical part of ecosystem studies. For example, the effort of scientists to bring back Atlantic Cod to the waters off the eastern coasts of North America has been a frustrating exercise in linear prodding of a complex system in an attempt to create abundance. Its failure reveals the limits of human ability to understand what conditions tip a system into a state of abundance or reduce it to less complexity from which it is impossible to recover.

INTERCONNECTEDNESS

As I have outlined it, three components contribute to illuminating the ecosystem of science by applying the complex systems approach:

1. Understanding the practitioners of science and their behavior as they go about their work—including the dynamics of interaction with other scientists and sources of scientific information.
2. Characterizing the interconnectedness and interaction as a part of the system, where the inherent behavior of scientists results in the patterns we see in science and scientific communication.
3. Understanding the communications networks that store and retrieve the information generated by the system.

As noted, each component within the system must exhibit the ability to adapt, compete, cooperate, and change (at least in some way) in order to survive and thrive within the complex system that is modern science.

The previous section discussed the role of the people within the system. From there, the next step is to examine the network of scientific communications. As I've mentioned, communication is akin to blood and connective tissue in the human body, or to the food web within the coastal ecosystem. Consider this: the discovery or addition to knowledge that makes any idea a scientific one is the communication of that idea in verbal, physical, material, or written form. The information being codified—literally put into "code" like writing or equations—becomes a tangible object that is measurable, transmittable, and actionable as science. Moreover, codified knowledge provides a feedback loop that also serves to "feed" the creativity of the scientific mind.

The knowledge communications system itself becomes worthy of consideration as part of the ecosystem of science. A number of researchers have noted that dense networks of communications characterize scientific research—as illustrated in the map of science. The relational nature of communications within research communities is particularly well suited to being studied as a network, since people will list co-authors, advisors, and other authors—a fact that philosophers of science have discussed for decades, but only now are we able to model in any satisfactory way. To rise to the top of their profession within science (we could say, peaks on a landscape), scientists publish the results of their research. It allows them to stake a claim on an intellectual landscape and to seek validation for their work. They communicate their work (at the most formal level, this is a scientific article) as they have done for more than 400 years. The communication of research, its processes, outputs, feedback, and peer review are all core features of scientific communications. These communications create networks among scientists—the networks have multiple roles such as exchanging resources, establishing norms, forming language (jargon), developing conditions for belonging and participation, and training for the next generation of researchers, all of which play critical roles in the scientific ecosystem.

Through scientific networks, scientists constantly react to the work of others. This reveals the interconnectedness of the system as connections are forming and reforming in response to changes in their environment—enabling adaptation and serving as gateways to different levels of the network. Changes can come from surprising findings such as that we discussed about the synthetic biology, as well as from social needs that demand scientific attention, such as that happened when HIV-AIDS emerged as a global health threat. In each case, the scientific community adapts and reorganizes in response to new "stimuli"—information within the system of knowledge creation. Some parts of the organization will remain stable—will be "conserved" in Kuhnian terms—as one would expect, and other parts change, sometimes wildly spinning into a new field of science. The newly evolved organization can be considered as an emergent property of the network of science, just as schools of fish are an emergent property of the organisms, environment, and feedback loops.

Finally, with regard to the evolution of science, scientific research shares with other complex systems a most interesting feature called *non-linearity*. One way to think about this is the contrast of spending a large amount of money (or time) on research that produces few results, for example, the ill-fated rush to announce "cold fusion," while a quiet field of research that is not garnering much attention or money that can suddenly produce an extraordinary, revolutionary finding, such as recent developments in graphene (noted earlier), which is to say in science the system's outputs are often not directly related to the inputs. Although some theorists have simplified the dialogue around science by describing an input–output relationship between funds and scientific knowledge, when viewed from the complex adaptive ecosystem perspective, it is possible to see that input–output models are not well suited to it. (And likely, even those who use such a description would admit its inadequacy to describe the relationships.) Indeed, it appears that the scientific knowledge creation and diffusion is non-linear—meaning the inputs are not directly related to the outputs and the whole is greater than the sum of the parts; small inputs can lead to large changes in the conditions of the system.

Individual components of the coral reef, if laid out on a dock, would not reveal the dynamics of the system as a whole. The emergent properties are non-linear—they can neither be reduced nor constructed. Similarly, scientific communications cannot be reduced to its component parts—scientific knowledge is an emergent phenomenon that results from the interactions of all the components.

While we have been likening scientific communications to a natural ecosystem, it is clear that human systems, unlike natural ecosystems, contain participants whose strategies in the science "ecosystem" are based on their interpretation (and reinterpretation) of past activities and anticipation of possible futures. These are features that are not found in the coral reef example. Agents in a social system are able to engage in intentional adaptive strategizing—they can intentionally make changes that contribute to the evolution of the system. This adds to the complexity of the system, which we'll discuss as we explore the global network in the next chapter.

REFERENCES

Doudna, J., & Sternberg, S. (2017). *A Crack in Creation: Gene Editing and the Unthinkable Power to Control Evolution.* New York: Houghton Mifflin Harcourt Publishing Company.

Gass, W. H. (2006). *A Temple of Texts.* New York: Knopf.

Hess, C., & Ostrom, E. (2005). A Framework for Analyzing the Knowledge Commons (Draft). In C. Hess & E. Ostrom (Eds.), *Understanding Knowledge as a Commons: From Theory to Practice.* Cambridge, MA: MIT Press.

Holland, J. H. (1995). *Hidden Order: How Adaptation Builds Complexity.* Reading: Perseus.

Kauffman, S. (1996). *At Home in the Universe: The Search for the Laws of Self-organization and Complexity.* Cary: Oxford University Press.

Kontopoulos, K. M. (2006). *The Logics of Social Structure* (Vol. 6). Cambridge: Cambridge University Press.

Monge, P. R., & Contractor, N. S. (2003). *Theories of Communication Networks. Computer* (Vol. 91). Retrieved from http://www.amazon.com/dp/0195160371

Padgett, J. F., & Powell, W. W. (2012). The Problem of Emergence. Chapter 1. In J. Padgett & W. Powell (Eds.), *The Emergence of Organizations and Markets* (pp. 1–29). Princeton: Princeton University Press.

Simon, H. (1962). The Architecture of Complexity. *Proceedings of the American Philosophical Society, 106*(6). Retrieved from http://link.springer.com/chapter/10.1007/978-1-4899-0718-9_31

CHAPTER 4

It's Who You Know (or Could Know) That Counts

Networks are the organizational feature of complex systems—the signature of complexity. Network analysis is central to fully understanding the patterns of communications that emerge within science and which represent the creation of knowledge. It is through networking that people achieve the vision of A.N. Whitehead (1925): "Modern science has imposed on humanity the necessity for wandering. Its progressive thought and its progressive technology make the transition through time, from generation to generation, a true migration into uncharted seas of adventure." The "wandering" takes place from person to person, connection to connection. In scientific communications, then, networks of co-authorships, citations, and project participation can reveal the underlying social relationships and allow us to trace the wanderings that lead to breakthroughs. The network is not flat nor random. This chapter describes these networks, their functioning within science, and what meaning they hold for participants.

When analysts study social networks, they do so with the understanding that the structure being examined—the nodes and links draw in two dimensions—is a *formalism* of the world, a set of self-generated relationships among actors or items. The structure is real—not a theoretical feature—but only part of the actual dynamic can be captured by analysis. To the analyst, networks represent the underlying social organizations and dynamics that operate in life, sometimes in ways that are not even understood by

© The Author(s) 2018　　　　　　　　　　　　　　　　　　　61
C. S. Wagner, *The Collaborative Era in Science*, Palgrave Advances
in the Economics of Innovation and Technology,
https://doi.org/10.1007/978-3-319-94986-4_4

the actors participating in them. Networks reveal less of the richness of personal contacts but more of the social construct than is known to and that influences the actors. Networks are organizations distinguishable from markets, firms, institutions, and families, and have different rules and norms associated with them. Within a social network, the knowledge, resources, or values that one communicates, represents, or offers—rather than one's titular position—establish one's influence. The ability to share a resource or reciprocate some value is critical to entrance and participation. This is a basic tenet of network theory: networks form to create opportunity for their participants. Position in the network is based on what each node has to offer to the network and what is taken up by other nodes—a researcher with highly relevant data may become more connected to others in the network because of the value of the resource offered. Networks also establish boundaries. People create and sustain connections largely because they become resources to others; *hubs* (highly connected nodes) take on the additional role of serving as hyper-linking resources within the network. These combined resources of nodes and links become a resource to those within the network. Using network-speak, connections are retained as long as they are of mutual (or potential) interest to participating members.

Participation in networks is fluid, which is part of what distinguishes them from hierarchies and institutions. Nodes may activate their membership or remain dormant based upon their needs and network conditions. At any point in time, a network is limited to resources attached to participating nodes and the structure provided by the links among them. Because they are fluid and changeable, networks quickly adapt to new stimuli or a change in environment, and they readjust or reorganize according to new information or the shift in resources. This makes network structures uniquely advantageous for researchers: as a new development, tool, or group member joins in, the network can readjust to take advantage of the new resource, and quickly communicate it to other members.

Networks have a relative independence and become resources as a form of organization within hierarchies, firms, bureaucracies, and markets. "Tapping into the network" is a common expression. Networks exist within hierarchies, even when unsought or unsupported. They do so by self-organizing around the interests of the members. Their features also make networks hard to study because they are quite different from command-and-control organizations, and input-output or cause-effect relationships, which have been intensively studied and are fairly well understood. We are less comfortable managing complex systems and their networks because we have had little time to learn about them. Only recently have scholars turned their attention to understanding complex systems and their networks. Advanced computing has provided the tools to study networks. Graph theory has advanced rapidly, but the associated fields have much ground still to cover to make sense of the network structures. Science stands to benefit greatly from understanding complex systems and their networks, if for no other reason than because it has operated as a network for four centuries.[1]

Grappling with complex systems has frustrated many managers and policymakers in part because people have a limited ability to deal with complexity. Thus far, the human tendency has been to stretch existing tools and mechanisms made for command-and-control system of management to try to manage complexity—with limited success. This tendency is rational because people tend to use the tools on hand, but the outcomes may not be as expected. Better we should examine the emerging system we are trying to manage. In doing so, we understand its basic structure and operating modes while asking the following question: what are the dynamics of networks?

NETWORKS ORGANIZE, OPERATE, AND EVOLVE

The collection of nodes and interactions as links determines the structure of what some scholars call the "typology" of the network—a word borrowed from cartography because of the similarity of concepts of a landscape. A two-dimensional geographic map shows some points as higher than others, some parts as flat, some sparse, others dense, some connected, and some not. In this case, think of a map of the geographic (not political) world, one that uses colors to show elevation and vegetation. Just as maps show commonalities across different landmasses, so too the layout of networks shows common typologies, sparse and dense landscapes, and connected and disconnected populations. Barzel and Barabási (2013) have pointed out that networks observed in nature and society share commonalities. The Worldwide Web network of linked web pages has features in common with, say, a network of film actor collaborations. Networks whose nodes do not seem to have anything in common can show non-random similarities, such as metabolic and transportation networks studied by

[1] This historic role of networking in science was the theme of my book, *The New Invisible College: Science for Development*, Brookings, 2008.

Guimera et al. (2007). This striking commonality of network structures suggests to many scholars that some laws of connectivity and principles of exchange underlie all complex communications networks, and that these principles and laws are knowable. Once known, opportunities exist for using networks to reach specific goals.

Networks of connected scientists have existed for centuries, as I discussed in my book, *The New Invisible College*. For now, let's zero in on the concept of network, considering it a "snapshot in time," so we can define the different parts and what each means.

Networks are usually visualized by identifying actors and using circles to indicate their presence (for nodes) and links (connecting lines) to show connections between them. In our case, the nodes are research practitioners (scientists, engineers, technicians, administrators) or possibly institutions (such as a university connected to a company or another university) or nations (France connected to Denmark). The lines between them indicate some kind of communication. The links indicate connections on multiple levels: for example, scientists and philosophers of past ages still "communicate" with us through their writings, which we can show by a link to a referenced article or book. We can show this temporality on a map: a *directed link* exists where node A is communicating to node B without node B reciprocating. Consider A as Einstein and B as the physics student in the twenty-first century: The student of physics may cite the work of Einstein without Einstein citing it back. This directed link is represented by a directional arrow on a network graph.

In most networks we will examine, the nodes share a two-way connection. Network analysts call these reciprocal relationships *undirected links,* meaning that nodes reciprocate some valued communication, with both nodes giving and receiving a communication of resource or value. "Reciprocal link" is a better term to describe these connections. The shared resource does not need to be equivalent in order to be reciprocal. Many times, research collaboration is not reciprocal or equivalent, such as when a senior scientist works with a junior scientist to train the other. The more senior person builds up social capital that can be claimed at some point in the future. Other times, resources are shared without the two nodes actually collaborating: simple cooperation can be an outcome of a link. Whether equivalent or not, in a network, a solid line between two nodes represents an undirected or *reciprocal link.*

Dense connections collect around nodes holding a strong reputation— this is the theory of preferential attachment. People who are connected are more likely to become more connected. Looking at the patterns of the network connections reveals a formal representation of the underlying social exchange, which includes people working at different points in their career. We mentioned training students. Senior scientists also connect to colleagues and to others with a similarly high reputation. As Harriet Zuckerman (1967) showed, the greater the reputation, the more sought after they are as collaborators, and therefore the choosier they can become about the quality of those with whom they work. In contrast, less experienced scholars (let's call them aspirants) do not have such a dense network of connections. This shows clearly in a network diagram where a "hub" will show many links, and an aspirant may be on the periphery of the network with just a few connections.

For now, by imagining a network of already connected scientists, we can see that information can travel from one node to another through existing nodes and links, just as you might expect to travel from one place to another on existing roads appearing on a map. I will use this roads analogy often in what follows, at the risk of stretching the point from time to time, because networks of roads are familiar, and their example can help us understand knowledge networks.

You are planning a journey by car. Even if the roads do not go directly to your desired location, your plans will take you on available roads toward your destination. You plot out a trip based upon available roads. If you are sightseeing, the trip along the established roads might produce a delightful turn where one can explore new vistas. If not sightseeing, then you might move quickly along the network of roads toward your final stop. The road networks offer a pathway of possibilities that are produced from the aggregation of many interconnected roads. There may be several ways to reach your destination. There also may be multiple modes to get there—car, bike, walking, and so on. Yet the mapped route also represents constraints. Travelers are limited to the roads existing on the map at that time. Perhaps the next time you travel this way, a new road will have been built, but for now, to move toward the goal you desire, you take the roads that have been paved. The trip is both enabled by the roads and constrained by them, in that other choices are not being offered.

A network of roadways is a limited analogy to scientific research networks, because roads are laid out on a more or less permanent basis and do not adapt to their environment, which human networks do. However, the idea of constraints on movement along the pathways is useful if we look at the network as a snapshot in time. It is possible to travel from place to place where the road links take you, and you generally do not go across space where the connections do not exist. In networks, if you are searching for someone who has information that you need, you begin by connecting with the people you know, who may be around you physically or socially. From there, a friend may point you to someone else as a resource, but generally, they will point you to the next useful friend. Similarly, during the learning or search process, a researcher is limited to the people already in their social sphere.

To continue with the roads analogy, if the desire is to go to a place that is not mapped, a traveler cannot tell from the map what's in the territory beyond the roads—it is unmapped and, therefore, unknown. A traveler needs some bravery to carve a new path into unknown territory. Consequently, only a rare traveler discovers a new path, in part because the "resources" that people need are found at the hubs—the towns or roadside stops. Food and water, or money and knowledge, are located at the hubs—important resources are rarely found "off the map."

Normal trips, like normal science, will travel along paved roads, stopping at established centers to gather resources needed to keep going. Goals are often known beforehand. We take the roads we know to take us to places we don't know.

The map of the scientific network similarly has established "roads"—connections between people, places, and ideas. Hubs (top people) and which connect the towns (say, a major university) and hamlets (a small college, a laboratory)—the less connected nodes—are well known. Hubs have many connections, but travel to less connected nodes is constrained to only a few or perhaps one connection. Similarly, in science, the network consists of those connections that can be documented in some way: people cooperate on a research project, they co-author a scientific paper, or they reference one another in some formal work—these communications are documented, and they can be mapped into a network of scientific collaborations. In science, researchers travel on established "roads" (and need resources from the nearby hubs), which may be the equivalent of Kuhn's idea of normal science. In order to innovate or challenge the existing paradigm, a practitioner must be willing to go "off-road" into the unmapped (or at least not well traveled) terrain, and they often desire to form a party of fellow travelers to go along with them.

The trip "off-road" is usually based upon a theory—hypothesis, estimation, guess –about what one will find in the unmapped territory. Perhaps the guess is based upon an observation of what one sees as an anomaly between the map and the natural world. It could also be based on extrapolating from observed changes in the terrain, as from plain to mountain. Existing maps may point to the possibility of undiscovered treasure, with tantalizing clues about what might be found in the unmapped terrain, but without providing clear directions as to how to get there. Perhaps a new road, or at least a new path, must be built. To get there, one might need just a simple instrument like a compass and one or two compatriots. However, it is also possible that the terrain is so challenging that a large engineering project is needed, requiring civil engineers and a road crew.

Again, the road analogy is limited, but it works to demonstrate the idea that some scientific exploration can be done on a small scale, while others, such as high-energy physics or space travel, may require considerable resources of diverse teams. To get at a new idea, for example, some previously unknown observation about quantum behavior, you may need to connect to some people not already in your network who know about this kind of behavior, or establish data connections with research labs that weren't previously involved, but which do relevant work. In between these extremes, there will be many other types of "terrain" to tackle with different-sized teams.

Several types of records exist to create maps of scientific communications. The most commonly used records are the names of authors on scientific papers. When two or more scientists co-author a paper, the names can be used to indicate nodes in a scientific communication network, assuming the relationship underlying the nodes and links. The network of communications needed to draw the map is evident from the names of people and their institutions listed in their publications. Other links are citations and references, acknowledgments, and keywords.

Networks Mean Collaboration

Over the past two decades, by every available measure, science has become more collaborative. There are more teams, cooperatives, collaborations, projects, and programs involving groups of scientists. In network terms, there are many more links established among scientists working at all levels of sciences. We can draw them as nodes on the map of science. Not surprisingly, more "nodes" are included, because of more people, institutions, and countries with scientists participating in the global network. Scientists now often form teams, which are inherently networks, to venture into unknown scientific territory. The team-building process takes place in three ways:

1. Governments announce the intention to sponsor additional research into new areas of research and call for collaborative proposals.
2. Scientists see an opportunity to advance knowledge across boundaries and organize a team of researchers.
3. An existing team spins off a new project.

The team-building process is usually one of *self-organization*—where researchers reach out across networks of connections to friends or friends-of-friends to seek people willing to join a collaboration to accomplish the next steps in their explorations.[2] The skills needed to move a project forward might not be known until researchers have begun to assess the exploration beyond the mapped territory and into unknown space. Researchers may simply not know what is needed to move forward until they have first traveled through the mapped territory to the edge of the unknown. Fellow travelers who know similar territory may know what is mapped and unmapped. Thus, an adaptable, adjustable, facile, and rich network of connections is highly valuable to those engaged in basic research.

NETWORKS OF RECORD

As the team develops and creates new knowledge, the practice in science is to publish advances in scientific journals. As work is formalized and published, the work itself becomes a link in another kind of network: a knowledge network. The publication event creates the opportunity to develop another types of mapping of the scientific terrain: the record of references and citations made by one scientist to the work of other scientists. Scientific articles list previous work (references) that they have used to support their own research. The previous work can also serve as nodes in a knowledge network, with links created by those scientific papers that cite them. The citation maps show intellectual precedence to one's research and place the scientific research into context—identifying the discipline within which the work is taking place and the prior work that the authors judge to be

---

[2] It is obvious that a researcher may also see the name of someone on a published article and reach out to that person to seek to collaborate with him or her, but we will discuss this type of connection later.

relevant. (Negative citations—references to work that is judged to be incorrect—are relatively rare, usually less than 5% of all citations.)

The mapped terrain of the knowledge network will be common to most trained scientists—the cited work represents the core of knowledge they know based upon their training and research. The territory adjacent to what is known, the *terra nova*, is the negative space carved out of the mapped territory. The adjacent space will also naturally be limited: the unknown space lies next to the known space. The ability to reach unknown territory is constrained by the pathways through the adjacent territory. This interdependence of the known and unknown territory may partly explain the reason, in the history of science, it is not uncommon for a discovery to be found by two people at one time[3]—the list of these occurrences is often cited as an odd feature of science. However, if we think about the unknown as adjacent possible space presenting the opportunity for discovery (and the existing knowledge and collaborative possibilities provide the constraint), then the next unknown space is not very random and, the next discovery, inevitable.

EVOLUTION OF COLLABORATIVE GROUPS

So far we have focused on a static view of the network to identify its components. Yet, networks are not static—in fact, their value to science is their flexibility and adaptability as they shift to accommodate the needs of research. Nathan Rosenberg (1982), in his book, *Inside the Black Box*, posits the best research assessment is to look at how knowledge is created. That means to examine closely and understand the dynamics of knowledge creation as it happens—from the initial conditions by which people connect, to the dynamics of a large research network—we want to learn how knowledge is created through networks. This can be useful to seeing inside the black box where communication and connection create knowledge.

While admittedly the literature on network structure, knowledge creation, and learning is in its early days, we can still celebrate that this line of research greatly advances the opportunity to understand science. In this regard, a social network is distinct from the knowledge network—the map of science. Network analysis can apply to both types of networks (social

---

[3] A classic article in the study of science, "Is Invention Inevitable," outlines many of the co-inventions in science. See, Ogburn, W. F., & Thomas, D. (1922). Are inventions inevitable? A note on social evolution. *Political Science Quarterly*, 37(1), 83–98.

and knowledge systems) because of the finding that networks follow underlying mathematical and statistical regularities, which expose the pathways open for information to travel. This allows us to examine social networks as having similarities. Social networks have different structures drawn from the distribution of links among researchers. In the map of science, the nodes can be research papers, and the links are connections made from papers to other papers based upon common words and phrases, references, titles, citations, institutions, disciplines, and concepts.

In order to create a social network of research practitioners, we look to scientific publications—which is the standard way that researcher practitioners report on their progress. Publications reporting the results of scientific research are widely used to measure science. Publication records are the artifacts that are more visible and enduring than any other kind of scientific communication; they are easier to count than other records that might be collected through survey or project participation lists. Publication records are also more permanent, and the lists of publications are widely available and reproducible by other researchers using the same methodology. Publications include articles, notes, letters, and reviews in an established, peer-reviewed journal or in conference proceedings.[4] Publications collected in databases[5] are the most used in large numbers to map the network of science, and within that, connections among authors as listed on papers are used.

How Scientific Networks Organize

While networks are used as tools in different types of analyses, in applying these measures to understand science, we need to examine the meaning of network structures for scientific knowledge creation: why and how researchers self-organize into groups, how collaboration operates, and what network typology can (and cannot) tell us about the human dynamics.

Let's start with the notion that networks self-organize, adapt, and change participants over time. Networks form around key people who are highly attractive to others for their knowledge, resources, or both. The

---

[4] The publication of scientific findings is undergoing significant changes as online services become more widely available. This proliferation of publication outlets will be discussed in Chap. 6.

[5] Two databases are most commonly used in science: the Web of Science (formerly called ISI/SCI) and Scopus. Many other databases exist, but they are usually dedicated to specific fields of science. For example, arXiv.net contains physics and computer science articles, while PubMed contains biomedical articles.

links that make a network are formed as scientific projects develop. They are assumed to remain in place after a project is finished—unless a person retires or dies. Even if links are not active, they remain potential resources that can be reactivated. Functionally, links develop upon a substantive relationship with a shared goal; in other words, the network represents working relationships, not a hallway conversation or an exchange of email. The network is more than the relationship between any two people: the professional links between people sharing a common goal, and the connections of these people with others, are a basic network. This means that the links between nodes in a social network indicate a collaborative endeavor that involve sharing and exchanging knowledge and support, and the potential to link to one another's connections.

The geographically dispersed collaborations and those that reach outside of one's own field of science are especially interesting because these links form (the members of the group select each other) even in the face of mitigating factors and obstacles. The varying environmental conditions such as funding, availability of equipment, support systems (students, lab space), and subject of inquiry all influence scientific organization. So let's address them before we talk about measuring networks.

Why do scientists self-organize into collaborative groups? The most obvious answer is that practitioners respond to incentives to take certain actions. Incentives both constrain and motivate action in ways that influence the scale and scope of collaboration. The incentives to connect have significant influence on the shape and structure of the resulting network. The incentives fall into four broad categories:

1. Equipment needed to make progress in research.
2. Location or difficulty of accessing/creating the studied object.
3. Conditions attached to research funding.
4. Location and accessibility of a desirable collaborator or resources.

The location of equipment needed to conduct research is one feature that heavily influences collaborative patterns. Some fields of science, for example, astronomy, high-energy physics, and seismology, require expensive equipment to conduct research—sometimes called "mega-science" centers. These large-scale instruments are built and then reside in only one or in a few places in the world. Equipment-based sciences, which funnel scientists together into a central location, require large funding commitments made over a long time; governments are the primary source of such funds. Because of the outsized investments needed, governments tend to organize the administration of such centers in a "top-down" fashion. Given the costs and intricacies of equipment construction, the effort to create a large-scale scientific facility needs a great deal of advance planning and coordination. These activities, such as the Large Hadron Collider in Geneva (Switzerland), have features of corporate hierarchies, with directors, managers, technicians, and laboratory technicians who support the scientific research. Networks still form within and beyond the institution, but overall, the institutional organization greatly influences the networks. It is clear that for large equipment-intensive research, scientific teams will be constrained in their progress by the location of equipment and the rules for accessing it.

In contrast to the mega-science centers, other objects of scientific study are scattered around the globe. Scientists study glaciers, coral reefs, geological formations, birds, or viruses, often in a unique place or at a specific point in time. Perhaps a collection (such as the ice core samples drawn from glaciers around the world) is kept in a specific place.[6] In many fields, practitioners travel to unique locations to observe a phenomenon or to access a resource. While the funding for the research is usually granted to a particular scientist, a principal investigator (rather than to the research site, as it might be in the case of equipment-based research), the location of the resource being studied requires that researchers or their teams travel to a particular place. Once on site, scientists report that they often meet other researchers, sometimes from other parts of the world, with research interests in common. Many times, these face-to-face encounters lead to collaborative research later on, or at least to data sharing and coordination of work down the line.

Still other fields of science such as mathematics or theoretical physics require only little or limited technology. On a spectrum of extremes with equipment requirements associated with high-energy physics at one end, to low-tech on the other end, the need for equipment (or not) has a strong influence on collaborative patterns in science. In equipment-based science, researchers will find themselves working together in order to access needed resources—often on a must-have basis. In areas that do not require extensive equipment, or that have other needs (such as access to a natural resource), the motivating factors for physical or collaborative contact will vary, and will usually be much less. Interestingly for our story, even in those fields that do not require large-scale equipment or access to natural resources, collaboration is increasing.

---

[6] My own institution, The Ohio State University, houses a world-famous collection of ice cores collected and curated by Lonnie and Ellen Thompson. The collection is located at the Byrd Polar Research Center in Columbus, Ohio.

DISTRIBUTED AND COLLECTIVE KNOWLEDGE

In contrast to equipment-based or resource-based science, many fields of science have experiments that occur across the globe connected only by electronic communications. Consider these projects as distributed research, in which an experiment is conducted across physical space and time, and in which different people take up varying tasks in order to contribute to research. This is important because it appears to be an approach to research that is growing the fastest of all. Consider the case of the Human Genome Project (1990–2005), where labs from around the world took on different parts of decoding the human genome, and then shared data in a common pool at the end of each day. Similar distributed tasking activities are growing quickly in number.

Social capital, the benefits of working together, creates the good will that makes communication and cooperation go smoothly. One method of creating social capital is by encouraging horizontal teaming arrangements that allow researchers to self-organize into groups. The team creates the meaning and structure within which cooperation occurs—this means developing a shared language and a sense of mission or goals, as well as developing a concept of belonging that will leverage and enhance the reach of a single person. The special needs of research teams was evident to the formative leaders of the Center for Nanophase Materials Science (CNMS) at Oak Ridge National Laboratories in Oak Ridge, Tennessee, a Department of Energy (DOE) laboratory, as they began operations in the early 2000s. The need for definition and belong was especially notable, in part because nobody had actually been trained in "nanoscience"—it did not even exist as a field before 1990. People trained in chemistry, computer science, medicine, and materials, all joined the team, and each had to take time to learn one another's "language" of research in order to organize. How long did this take? "Six months," answered one chemist definitively, to which others on his team nodded their heads in agreement.

With the various kinds of resources provided to scientists, networks are highly efficient when it comes to creating knowledge, but they do not *retain* knowledge well, and neither are they effective in making decisions. When decisions need to be made, leadership is critical. When knowledge needs to be retained, codification is needed—it must be written down or built into equipment; and institutions are the place to hold it. Consequently, even in a networked world, science must maintain institutions to retain knowledge. That is where it is held among people who share a worldview, along with a specialized vocabulary that has been developed for their discipline. Institutions serve the important role of retaining knowledge gathered into a discipline.

The DOE nanoscale science centers established in the early 2000s invested in expensive, unique, high-tech equipment. This made them, the five centers including CNMS, attractive hubs for researchers. But the directors noted that, given the pace of change at the cutting-edge, it was only possible to plan a few years ahead for the equipment-based needs of research. Lab directors do not know in advance the size, scale, and scope of the team or the equipment needed to make advances. This part of the scientific process must self-organize and evolve as knowledge advances. The team then is able to feed back into research planning the kinds of resources needed for additional phases of work.

Individual scientific researchers, who often act as self-directed agents, asking questions about the outer edges of knowledge, use networks to gather information, identify collaborators, and exchange questions and capabilities. Scientists seek out these connections as new questions arise. History shows that researchers have responded to new challenges in this way since the rise of the earliest experimental science in Europe in the seventeenth century and later in other parts of the world. Sociologist Diana Crane, continuing earlier work discussing science as a network, found among scientists the desire for originality as the motivating factor for creating and maintaining contacts with scientists in fields different from their own. For most scientists, the desire to be original—novel, creative, path-forming—requires knowing what has been done or what research is already underway. The bridging function across otherwise unconnected fields or clusters in a network is often related to originality and the emergence of new ideas.

This bridging function to spur knowledge creation across disparate sources of knowledge and information was a key motivator or incentive for the directors of the nanoscale science labs, too. The goal was to create a center where a core group of top researchers would be working and available as resources to others. Researchers from around the world would come to work with the CNMS scientists, as much as with the world-class equipment at the DOE labs—this was the original design for the centers. CNMS scientists would benefit in a positive feedback loop from connections with top researchers from outside the lab. The designers intended that the CNMS and its sister centers would be *collaborative centers* involving center staff working among themselves, as well as welcoming external researchers from as far away as China. External researchers are not funded by DOE, but the intention was to attract them to the unique collection of resources.

In contrast to past practice, within CNMS, security restrictions on visitors were relaxed. Visiting the center for the sole purpose of using equipment was discouraged, since the emphasis was on collaboration, exchange, and networking. Additionally, external collaborators were asked to join onto a project, or to have a basic research question in mind when they applied to work with people at the center. Collaboration was built into the structure of the CNMS from the start, and visitors had to show that their research question was suited to the equipment and staff available at the DOE nanoscale center before they were welcomed into a collaborative group.

LIMITATIONS TO NETWORK COLLABORATION

Belonging to networks has many advantages, and researchers often strive to be included in networks. However, choosing to create or strengthen a link has a cost; the nanoscale science centers provide a good example of how this works. Researchers from with the federal lab or at a local university who joined the CNMS had to forgo other opportunities to conduct research in other places. The commitment required to join a long-term project meant making a clear choice to be on the CNMS team. It is a full-time, long-term commitment.

Intuitively and in general, each person is limited in the attention they have to give to social connections at any one time—a fact that has been studied and quantified within a number of disciplines, including information science. For example, there is Robin Dunbar's number,[7] where the anthropologist suggests a cognitive limit to the number of people with whom one can maintain stable social relationships. Dunbar predicted a human "mean group size" of 148 (casually rounded to 150), a result he considered exploratory due to the large error measure (a 95% confidence interval of 100–230). Similarly, philosopher of science Derek deSolla Price suggested that a subfield of science consists of 200 people. (Beyond that number, Price suggested that the subfield would branch off into yet other subfields.)

Intuitively we know that human attention is subject to limitations. Once a person has chosen to work with someone else on a team or at a lab,

---

[7] Dunbar, R. (1992) Neocortex size as a constraint on group size in primates, *Journal of Human Evolution* 20: 469–93.

it implies they have passively decided not to work with someone else—often meaning other links become dormant. Moreover, when choosing with whom to work, a person often connects with the friend-of-a-friend (a small-world connection), adding an initial trust component to a collaboration. In this way, everybody (including scientists) "filters" their network connections to a manageable size.

An individual is also constrained by the extent to which he or she can lay claim to and offer resources to others, since membership and position in a network depend upon reciprocity. A well-connected node (say, a scientist with coveted access to an expensive piece of equipment) has resources to share: this gives them additional influence within the network. The well-resourced node (perhaps the one with a large grant of funding) is positioned to be more influential than a less-resourced node. The well-connected/resourced person is often able to choose with whom to work: they become a broker—a node with power to include or exclude others.

THE PERSONAL DYNAMICS WITHIN NETWORKS

The conditions within networks, such as how easily knowledge flows, are conduits for influence. As new members arrive in a network, they often look for a link to someone more powerful or better connected than they are. This process of seeking to connect to someone already connected is called *preferential attachment*. It is a network dynamic that affects evolution of the network as new members enter into collaborations. Newly arriving members must have something to "trade"—such as knowledge or time. Connected members may accept a new link from an aspirant in exchange for a resource (perhaps the aspirant can conduct specific tasks). To the extent that the existing member can pay "attention" (à la Dunbar's number) to the new entrant, they may welcome a link. These are largely voluntary connections from one voluntary member to another, creating constraints and affording opportunities to those practitioners in the network, and frequently changing the structure of the network.

The nature of the links among scientists—their connections and reasons for organizing into a groups, clusters, and networks—is established in science as it relates to two things: (1) the kinds of problems that are being studied and (2) the rules of affiliation. Studying scientific co-authorship networks in 2000, University of Michigan mathematician Mark Newman found that scientific networks tend to present a structure of densely linked clusters connected by a few weak ties. Visualize this as tightly connected groups of people, speaking the common vocabulary of their discipline and sharing certain resources with a few people (the weak ties) experimenting with interdisciplinary or boundary-spanning ideas with other groups of people (clusters). This structure may have been unique to the field studied—and this question about expectations for network typology becomes very important to our story.

The structure—the typology—of scientific collaboration networks can vary considerably based on the maturity of the network, the nature of the problem being explored, the equipment needed, the location of the resource being studied, and whether the communication needed is suitable for asynchronous connection, distance links, or the opportunity to evolve over time. Mark Newman observed that networks dominated by many people with a few collaborators (rather than a few people with many connections) characterize biomedical research. It is possible to see that other scientific networks are characterized differently. For example, in a study of basic research in Europe, I found that such research is characterized by dense interconnections more than clusters. This suggests that basic research may have more open sharing than, say, biomedical research, where results may be translated into marketable products.

An evolving network, one focused on creating new knowledge and establishing research parameters, can be expected to change shape rapidly. Basic research networks can be expected to be denser and less clustered as people seek to retain many connections and allow information to flow freely. There may be more preferential attachment going on within dynamic networks, as people seek to make connections with researchers having greater reputation or more resources, typically senior researchers. As senior researchers become maximally connected, aspirants will need to look to other places for connections. A less mature research area may reveal a denser typology that may change rapidly, where a more mature research field can become fixed while exploring a research question. This maturity shows up in the network as a cluster of researchers when the collaboration is efficiently using shared resources.

SHAPE AND STRUCTURE OF THE GLOBAL NETWORK

What is the relationship between network typography and dynamics, and the maturity of a field? How long has a group of, say, 200–400, people seen themselves as constituting a subfield of science? How open are they to newcomers? How many boundary-spanning ties do they have and can we 'see' them for the purposes of supporting new developments?

If we consider science at the level of a "discipline" or "field"—those researchers having a central problem with items considered to be facts relevant to that problem, and having explanations, goals, and theories related to the problem—it is possible to establish an expectation about the 'shape' of the network. Subfields with accepted paradigms can be expected to have a network structure that might be highly stable and clustered into teams. Indeed, this is what we find when we look at the established fields of science. They show high stability and high clustering in groups that correlate with an underlying accepted paradigm.

The disciplinary structure of science—the categorization of science as "physics," "chemistry," "genetics"—is well known and widely accepted. Many realize it as a stable artifact of nineteenth- and twentieth-century social and political organization that influenced the growth of science. Sociologist Julie Thompson Klein[8] notes that the modern concept of a scientific "discipline" has been in common use for only about a century, emerging at the end of the nineteenth century.[9] Prior to that time, distinctions defining disciplines had little meaning to practitioners and experimentalists who worked toward developing what they called experimental philosophy. Coincident with the advent of scientific societies in the seventeenth century, members did not differentiate among types of disciples except in the broadest sense, such as "medicine," "natural physics," and "mathematics."[10] Disciplines evolved slowly in the eighteenth century, and then more quickly in the nineteenth century as dedicated funds began supporting the professional laboratories of Pasteur and the Curies.[11] Even at their most stable (disciplines in the early twentieth century were very clearly delineated), networks continued to operate, although with less fluidity than we find now. When disciplines reign, institutions are more prominent as they hold the historical record and serve to retain (conserve)

[8] Klein, J. T. (1996). *Crossing Boundaries: Knowledge, Disciplinarities, and Interdisciplinarities.* Charlottesville, VA, University Press of Virginia.

[9] Martha Ornstein (1928), *Rôle of Scientific Societies in the Seventeenth Century* (Reprint edition: Archon Books, Hamden & London: 1963) the *Philosophical Transactions of the Royal Society* of London is the first modern scientific journal The terms "science" and "scientist" did not come into common use until the nineteenth century.

[10] D. D. Beaver and R. Rosen (1978a): "Studies in Scientific Collaboration. Part I: The Professional Origins of Scientific Co-Authorship," Scientometrics, 1, 65–84.

D. D. Beaver and R. Rosen (1978b): "Studies in Scientific Collaboration. Part II: Scientific Co-Authorship, Research Productivity, and Visibility in the French Elite," Scientometrics, 1, 133–149.

[11] Ibid.

knowledge. Encouraging new knowledge creation may have been part of their mission, but a far less important one, to be sure, than their role in retaining knowledge.

So how does new knowledge emerge within these disciplines? One answer, to use the language of networking, is that new knowledge emerges from the action of those who reach out of their historical disciplinary clusters in an exploratory process that spans boundaries. Susan Mohrman, an organizational development scholar at University of Southern California, studied how the subfield of Massive Parallel Processing (MPP) emerged from the fields of mathematics and computer science. Her research revealed the growth of underlying networks that formed as people from different fields found they needed to work together on modeling and simulation capabilities. Like the nanoscale science research, MPP was an emerging field in the 1990s. The initial problems that needed solving demanded that practitioners seek out new knowledge from people outside their field. They needed to recombine knowledge by integrating ideas, theories, and capabilities in a novel manner, drawn from various other disciplines, in order to reach the goal set out by the initial question. The initial question arose less from an anomaly in data than from the opportunities posed by new computer hardware. An exploratory phase was followed in the research process by the formation of a cluster of practitioners. This would look, in network terms, like a sparse network becoming denser, and then becoming more clustered. A visualization of the network would look much more stable as the group sought to exploit the social capital created by the forged connections—a process that in the MPP case took place over many months. This was followed by a more focused phase of development within the emerging paradigm in which a newly formed subfield began working on modeling and simulation. The number of adherents to the concepts and experiments grew, and a common language of collaboration was developed out the early MPP work. A subfield became established that had a clear and relatively stable network typology based upon the connections created in the research phase.

Professor Mohrman found that, at some point in the research process, the researchers defined certain problems that could be solved within the group; as they solved the problems, the process became routine. The group had forged a community with a common understanding; this in turn enabled them to define research that was even more complex. As order stabilized at each level—just as might happen in the ecosystem when order is established, say, at the level of coral reef—it enabled further diversity.

Another way to say this is that it enabled a greater degree of freedom. In the MPP case, this ordered state at one level and the definition of new avenues for research led to the search for mathematical breakthroughs that could establish a platform at the next level for the generation of simulation-based research. While this was going on, the team reorganized, which allowed new entrants to join and saw some members depart the team. In this way, linkages were again formed to introduce variety (exploration) rather than continuity (exploitation).

Research shows that the relative presence in the MPP team of coopera-tion versus competition also changed through time as a research team went through its various stages. At first, a new cluster or niche tends to be "open" in that there are resources for early entrants to seek and exploit new ideas: this is shown in the density of connections in the early network. There are relatively few contenders to compete with in the early stages—cooperation is highly adaptive and useful at this stage. As a niche or sub-discipline emerges and becomes more heavily populated, competition for resources develops—this is reflected in greater preferential attachment, more competition, where hubs develop and competition for links to more reputed members becomes important. This process leads to an interest among the members to reach out across their networks and to strengthen affiliations of various kinds. The teams become more attractive; clusters of mutually cooperative relationships grow, competitiveness can begin to grow across clusters. This dynamic can lead to the establishment of various symbiotic relationships, based upon mutual or complementary interests. Complexity increases, which enables further diversity, and community emerges. At this point in the development of a scientific niche, the subfield becomes defined enough to allow the establishment of entities like CNMS—the nanoscale science centers described earlier. Such centers can support coordinated approaches to securing resources and enable the clus-tering of knowledge, people, instrumentation, and funding. A subfield has fully emerged.

MEASURING NETWORKING AND GROUP DEVELOPMENT

Network analysis can measure the level of a group's development and, along with that, define expectations of network connections. Different expectations can be detailed at each level based upon the understanding of the need for a group to find new ideas and connections—with the expectation of a collaborative process at the formative level, and then the acceptance of ideas into a new ordered level (such a subdiscipline), which then becomes more competitive. Cooperation proceeds competitiveness. At some point, the new ordered level establishes a parameter of organization as a new subfield, often heralded by establishing a new science journal such as *Nano Letters*, established in 1995.

Each level of group organization has an expected network typology of communications dynamics. Connectivity measures of a network can reveal the extent to which new entrants are able to join the network and whether preferential attachment is operating (this will reveal the openness of the network). Measures can also reveal the extent to which the members are linking to one another in dense subgroups (lots of cross connections) (cooperative) or in clusters (groups densely connected to one another but only weakly connected to other groups) (competitive). The extent to which new members are joining the network can be measured by counts of new members (or nodes) being added to the network. Several possible measures can reveal the extent to which members link to one another once they are in the network. One of these measures is transitivity—a measure of the extent to which any two nodes connected to a hub are also connected to each other. This measure focuses on counts of the links between nodes at the periphery as they relate to each other. It is a useful measure in science because links to the nodes are obvious, but connections at the periphery are often only latent connections or possibilities that are not realized at the time of measurement.

A second measure of connectivity is called *mean degree centrality*—a measure of the extent to which a network is centralized around powerful members. A highly centralized network concentrates power and influence among the core members, making it difficult for peripheral members to join or access information within the core. A more established field is more likely to have this typology. This is a well-known social dynamic to any teenager who wants to be part of the in-crowd. The "popular kids" often control who is in-and-out of a group by clustering friendships and shunning others. Of course, scientists are not (usually) this crass about choosing collaborators, but the idea of centralized network power is often experienced within science, too, as when a well-reputed scientist, already saturated with connections, is approached by a junior researcher.

To examine the extent to which the Department of Energy was investing in open research networks at the beginning of their nanoscale science endeavor, measurement was made of the nanoscale science networks around the DOE labs. These measurements showed that the nanoscale

networks at the time the CNMS was established were not highly centralized. This was expected since the field was new and growing rapidly.
Therefore, analysts expected to see dense connections across a broad network. For DOE, this network structure was good news, since the network
was not clustering around a few powerful and attractive researchers. From
the measurements I made, we deduced that knowledge could flow relatively unencumbered through the network—as one would hope to find in
a basic research network where a great deal of exploration takes place.
Further, it meant that connections between nodes could be made (and
broken) fairly easily because many connections were possible, adding to the
potential efficiency with which effective exploratory teams could organize.

In network analysis, a measure called the *clustering coefficient* measures
the extent to which nodes are redundantly linked to each other. Here,
clustering represents groups that form into a goal-oriented team that cul-
tivates common knowledge and a group identity, often to the exclusion of
others in the network. This development in the network generally indicates that subgroups are forming to combine knowledge (often by forming a common language) in search of a common understanding and
problem-solving methods. In Kuhnian terms, they may be in the process
of challenging the existing paradigm in search of new explanations for
phenomena that cannot be explained by current knowledge, but doing so
as a cooperative group rather than a sole explorer. Greater clustering
would be expected in more established fields.

I observed this development of groups (clusters) in the networks associated with CNMS. At the start, the networks were dense, but they began
to grow more clustered over time as the wider group formed into teams.
The teams began to define goals and distribute tasks among their members. Members of the team reciprocated with resources shared among
themselves, offering valuable assets in exchange for complementary assets
from others. The teams began to compete with one another for resources.
The competition spurred excellence as the teams sought to define and
distinguish themselves for access to resources—and to make themselves
attractive to outside collaborators.

With these network analytical tools in hand—clustering, density, and
centralization—we can watch networks grow from sparse to dense to clustered, and we can see how easily (or not) knowledge can be accessed and
can flow across the knowledge network. Using just these few tools from
network analysis, we can look back at the complex ecosystem in the aggregate, or at some ordered level, where measures reveal striking insights into

the dynamics of the emergence of the ability of groups to create and communicate knowledge.

Taken together, the scientific ecosystem I've been describing through network analysis and its associated terminology reveals a diversity of nested structures in a heterarchical organization. In a knowledge system, we expect that some parts will be dense (enabling exploration) and other parts may be clustered (enabling exploitation). These patterns are not random, but neither could they have been anticipated by studying individual participants. The properties of the system at the aggregate level—the emergent properties shown in teams—become a resource to those at one's level and to the next higher level. We can readily see that no single actor in the network is responsible for the properties of the network, whose resources eventually become of great value to those in the group. Neither are the properties constructed according to a plan: the knowledge system emerged from the activities of all the agents in the group, who act according to rules, such as preferential attachment or reciprocity that characterize motivations and behaviors in the system. Scientists have been described, for example, as engaging in ongoing social exchange to carry out their strategies and as staking their reputation by collaborating with others in pursuit of their personal objectives. Indeed, that is the case, but their actions at any one time are constrained by the structure of the team (representing the knowledge system) to which they are linked. Such collaboration may end up contributing to the larger goals of the community—such as having CNMS create a "lab on a chip" with nanoscale particles. It also helps sustain the emergent properties in that the network in which a group of highly trained researchers has developed additional capabilities. The resources that are exchanged often reside in no organization or intermediate formal arrangement (such as a formal alliance)—nor in the market; they reside and are evidenced by the self-organizing community we call a network.

IMPLICATIONS OF GLOBAL SCIENCE AS A NETWORK

This interpretation of science as a network—while not a new idea to science studies—has been energized by recent advances in both computing and algorithm development. This advance has attracted analysts from far-flung fields (physics, sociology, information sciences, biology) to examine the interconnections within scientific research (primarily as text) in order to view it in network form and study its dynamics. Some of these analysts have been attracted to the study of science because of the huge, high quality, historical publication databases that contain records of publications in journals of scientific research. The databases, for example, Thomson-Reuter's product, Web of Science, enable rich, time-series, and text-based analysis. This work has provided many of the insights into network dynamics we can use to monitor the global network.

As shown in the MPP and the nanoscale science cases, scientific networks form around problems—by including the right participants, coming to the point of scientific exploration, by traveling along the established pathways (theories/paradigms), and then seeking to access the adjacent possible space—researchers seek a path into the unmapped territory. In this process, groups of researchers exist at multiple levels of inclusion, and as overlapping and nested sets—individuals can belong to multiple communities (with a limit) and can have cross-links to other communities. Each link can be activated or left latent based upon constraints and opportunities.

An important characteristic of scientific networks is their fluidity, in terms of both participants who come and go, and linkages that are established and broken. People join or leave groups for many reasons, only some of which have to do with the network. Within the network, scientists also wish to "survive and thrive" just as any agent on a landscape. The behavior of scientists is driven not only by a particular problem they are trying to solve but also by their personal strategies and the networks to which they have access, all within the larger science system and the opportunities it provides. As I take pains to point out, this is not a random process: it follows patterns and processes. This allows us to anticipate the dynamics at each nested level: the individual motivation, the newly forming team, the established team, the subdiscipline, and the formal disciplines, all of which interact within the heterarchical structure.

It is important to understand that for collaboration in science, the principle of self-organization can explain the emergence of complex forms of cooperation at each level where external structures and plans cannot. To put it one way, "You can't legislate scientific discovery." In fact, some have argued that complex activities are inevitably self-organizing—that they cannot be fully controlled, externally or hierarchically. Within self-organizing systems, such as the ones that constitute the heterarchy of science, the emergence of order cannot be attributed solely to top-down or bottom-up dynamics, as we have seen. On the one hand, the leaders of the nanoscale science centers did not direct person A to work with person B—the researchers themselves made that decision. On the other hand, the bottom-up connections were partly determined by the interests and capabilities of those in the network—while the organization was neither top-down nor bottom-up, the opportunities were constructed by planners and policymakers. That process involved leadership and conservation, policy and decisionmaking, as well as exploration and exploitation. Order may appear to arise spontaneously from local interactions of participants even if they are not aware of how their actions contribute to a larger order; although if we look closely, we see the constraints that guided the exploration to unknown territories just off the mapped terrain. We will discuss the constraints in more detail.

EXPLOITABLE CHARACTERISTICS OF SCIENTIFIC NETWORKS

Within the networks examined so far, the four conditions for self-organizing systems in general, and knowledge systems in particular, are met:

1. They are organic and grow from within, and they renew independently of direction from outside the team.
2. There is mutual causality in that people within the network have shared (although not equal) interest in creating connections.
3. There is "requisite variety" to address the problems of interest, and as the system attracts new capabilities, the variety of knowledge available increases.
4. The system is not in equilibrium, but rather is in a transitional state that provides the opportunities for self-organization and the generation of further emergent properties.

REFERENCES

Barzel, B., & Barabási, A.-L. (2013, September). Universality in Network Dynamics. *Nature Physics, 9.* https://doi.org/10.1038/nphys2741

Guimera, R., Sales-Pardo, M., & Amaral, L. A. (2007). Classes of Complex Networks Defined by Role-to-Role Connectivity Profiles. *Nature Physics, 3*(1), 63.

Ornstein, M. (1928). *Role of Scientific Societies in the Seventeenth Century.* Hamden/London: Archon Books (Reprint Edition, 1963).

Rosenberg, N. (1982). *Inside the Black Box: Technology and Economics.* Cambridge: Cambridge University Press.

Whitehead, A. N. (1925). *Science and the Modern World.* New York: New America Library, Macmillan.

Zuckerman, H. (1967). Nobel Laureates in Science: Patterns of Productivity, Collaboration, and Authorship. *American Sociological Review, 32,* 391–403.

CHAPTER 5

The Global Network of Science Emerges

Maps of the political world became obsolete in the late 1980s. The reunification of Germany and the breakup of the Soviet Union changed not only the political map but also the possibilities for sharing science and technology. For the decade prior, the European Union (EU) had been conducting a social experiment in regional consolidation of the research community, which also influenced the state of research and development. In Asia, South Korea's science and technology system had reached first-world levels of capacity. Other Asian governments also showed enhanced interest in science and technology through government policy, investment, and corporate commitment to knowledge-based industries. As the politically motivated Cold War competition between the East and the West ended, science itself was the winner, as collaboration began to grow at a global scale; duplicative capabilities in different parts of the world could be turned toward coordination, cooperation, and eventually collaboration at the frontiers of knowledge.

The technological world also reached a turning point at the end of the 1980s. Arpanet, the computer network created with funds from the United States Department of Defense, was reconstituted into NSFnet to assist the scientific community with rapid communications. The computer-assisted knowledge network had been born. The earliest applications of computer-assisted file sharing emerged with applications built by clever minds in

© The Author(s) 2018
C. S. Wagner, *The Collaborative Era in Science*, Palgrave Advances
in the Economics of Innovation and Technology,
https://doi.org/10.1007/978-3-319-94986-4_5

computer science. Eventually, one of those minds, Tim Berners-Lee, working at CERN, developed the World Wide Web application. This and other public research investments, ones that had supported national security research, began to be repurposed toward large-scale scientific and technological research projects that had taken a backseat to defense research. The political changes released the pent-up energy of scientists, engineers, and policymakers to reach beyond the boundaries and borders that had defined the post-war and Cold War eras. Science and technology projects that had serviced national goals could now be turned to the frontiers of science and technology itself. (Similarly, the computer capacities increased to enable the analysis of large-scale network data, such as these.)

As scientists and engineers began reaching beyond the politically enforced boundaries of the post-war era, the early 1990s saw the birth of the global era for science and technology: a network had emerged to become the most influential force and dominant form of organization for the knowledge-creating community. The global network now encompasses every country in the world (even North Korea can be seen among the peripheral nodes) and determines the agenda for those forms of organization lying below it.

Scientists and engineers—released from the constraints of national borders—began to create connections with one another beyond institutional boundaries and across national borders at rapid rates. In addition to sheer growth—which has been spectacular in itself—the network has developed characteristics that emerged from the interactions of thousands of practitioners, creating pathways and connections that built opportunities for other linkages that followed. This chapter describes these characteristics.

Large-Scale Properties of the Global Network

The large-scale statistical properties of the global network reveal both the dynamics of interactions and the pathways that created knowledge. The established pathways build bridges, launching the potential for subsequent linkages for those who come later to build upon the earlier connections. The network has grown at a rapid rate by the addition of "nodes" (addresses) and "edges" (links represented by co-authorships). In 1990, addresses in the Web of Science database revealed 172 country names on internationally co-authored scientific papers. By 2013, this had grown to 230 country names represented in the international network—nearly every country in the world.

The number of countries participating in the global network increased linearly from 1990 to 2013, but the number of co-authorship links among these countries more than tripled. The average number of links per country was 11 in 1990; that grew to more than 36 by 2013. Simple division gives us the average number of links per node, but averages do not mean much in the global network, because, on size alone, larger countries will have more links. The United States, the EU countries, Japan, and Switzerland will have more collaborative projects than smaller, less-wealthy nations. The network data were normalized for the size of the countries, but even after this calculation, the larger, wealthier countries simply drew more collaborators.

The network has developed into a stable structure. It is barely perturbed by the comings and goings of individual agents or projects. The network measures show that the global has become denser (more overall numbers of links). Another way to examine density is to measure the average degree, which is the number of links per node, averaged across the network. The network's average degree has risen from 22 in 1990 to nearly 75 in 2013, suggesting that nodes in the network are greatly increasing their connectivity. If we think of nodes as stepping stones, the average distance across the network is just two steps—an astonishingly low number—suggesting that the density of the nodes and links create the opportunity for those within the network to be just a few "handshakes" away from one another.

Over the 13 years studied, the average degree has risen, further confirming the tripling of links among nodes, with the acknowledgment that the "richer" nodes have surely attracted many more connections than the "poorer" nodes, which have joined into the group, but perhaps not as tightly knitted into the fold. The average degree shows more than the growth in links among existing nodes: it shows that the network is becoming more resilient, meaning that changes in configuration of connections among participants are less and less likely to affect the structure of the network as a whole.

The network measure of betweenness centrality dropped over the study period. This indicates an interesting and potentially important finding: influence within the network actually became more dispersed over time, rather than more concentrated. The more central nodes are considered to have more influence on flows of information across the network. Network theorists have said that central nodes could distort transmission of knowledge by withholding it, distorting it, or favoring some over others. Some have questioned whether the global network is reconstructing political relationships, but this measure would suggest the opposite. Centrality is dropping, suggesting that more and more nodes are acting as communication hubs. This reduces the opportunities for distortion or favoritism, perhaps improving equity and reciprocity.

The clustering of nodes in the network is remarkably high, and it grows slightly over the years examined. A *cluster* is a fully connected triplet of nodes. A *clustering coefficient* measures the extent to which nodes across the network appear as triplets. On the one hand, the clustering coefficient adds to the interpretation of shared power across the network, because lower clustering could indicate strong hubs which could hold a great deal of power and reduce efficiencies of communications flows. The high clustering coefficient shows that most nodes are parts of groups, not satellites of large hubs. On the other hand, clusters can sometimes make it difficult for new members to join the network since it is harder to break into a cluster than attaching to a single node.

SMALL-SCALE PROPERTIES WITHIN THE GLOBAL SYSTEM

The global network grew—at least in the early 1990s—as individual scientists chose to link together. As the network expanded and stabilized, the more elite the scientist, the more likely he or she is working at the global level. Well-known scientists tend to work with each other across institutional and political boundaries. This makes sense in a self-organizing networked environment operating under rules of preferential attachment. As a scientist becomes better known, more people want to work with him or her. As more people contact the rising star or elite scientists, the star can be more selective when choosing with whom to cooperate. As an experiment is designed, a researcher chooses among collaborators; she or he will want to choose one who complements her or his own knowledge, as well as one who is not going to add to social obligations beyond those to the science itself. The reputation of the collaborator will also be important, particularly if the two researchers have not worked together in the past. Efficiency, complementarity, and reputation will be important in forming a team.

In addition, the more well-known scientists become in their chosen areas of research, the more likely it is that others from far afield will seek to connect. As this happens, new ideas are shared and spread, even if a formal collaboration does not result from a communication. Geographically distant collaborators can offer significant benefits in terms of reduced social demands as compared to people who are nearby. Geographical distance can actually improve outcomes and enhance efficiency because it can reduce social obligations. More subtly, the research project does not need to continue past its usefulness to both parties. In short, international research projects can be easier to start and certainly easier to dissolve than dealing with close associates. As a result, distance can actually increase efficiency, and distant connection can provide high scientific value in terms of diversity and creativity.

Indeed, we often see that research networks grow with the least social encumbrance at the global level. Social and administrative pressures that might impinge on people working within a physical laboratory are absent at the global level. There is less need to navigate personal relationships in forming a collaboration. There may be no lab director or branch chief to meddle with team organization. No global ministry of science coordinates practices at that level. In sum, there are few if any political pressures present to bias choices for collaborators. Very few organizations offer funding for globally collaborative teams. With the exception of the EU, few government agencies offer funding for international cooperation. The motivations for connecting at the global level—in early days—are almost all prestige driven. This dynamic drives up the value of a researcher's time, and thus it increases the likelihood that the rising star or elite scientist will choose to work across international lines—usually with other elite scientists similarly pushed up the vortex of recognition and prestige by the dynamics of preferential attachment.

International collaboration in science presents a dilemma. At the same time that working at the global level offers freedom, there are countervailing forces that present a number of obstacles to global collaboration. These include time zone differences, language barriers, divergent norms and practices among institutions and cultures, obstacles to sharing funds, and perhaps unfriendly peer reviewers who are less inclined to recognize international work. Intuitively, these and other factors would appear to be working against international collaboration in science. One could draw the conclusion that scientists working together at the global level are making a conscious choice to buck the obstacles in order to forge a partnership. And, they are making a freer, purer choice to collaborate than might be made at other levels, such as within lab, within discipline, or within country. More countries of the world are involved in science at the global level than at any time in history, so it is clear that this activity is offering benefits.

This makes the global network of scientific collaboration an extremely interesting phenomenon. Why has it grown so fast? Perhaps it's because the global level offers a degree of freedom within a scientific ecosystem not found anywhere else. Many top-level scientists are working at this level. Many unique resources are available to the researcher at this level (equipment, natural phenomena), and access to them occurs with fewer constraints found at other levels. If we analogize the lower orders of scientific communication to levels of the ecosystem, then we might say that the global level offers the opportunity for maximum creativity and diversity, because opportunities are unencumbered by institutional requirements.

SETTING THE GLOBAL AGENDA

As it has grown in members and link density, the global network has taken on its own organization with norms, rules, and patterns of information sharing. Papers produced by those at the international level are more highly cited than sole-authored publications. Notable scientists are likely to be working in the global network. The stabilization of the global network could mean that global participants will set the agenda for the national and local levels. Describing the influences in ecosystem and network terms can help us understand the influence of the global network on other systems.

While the global level appears to be beyond the reach and influence of most national governments, the networked structure of global communications provides opportunities and benefits that are not available at the national level. Learning to tap these benefits is the challenge posed by the rise of the global network.

The growth of global science has occurred in sync with other forms of collaboration, a trend toward interdisciplinary research, and team science. Up until the early 1980s, most scientific papers had domestic authorship, and the tradition was to publish a paper with a single author (even when cooperation may have been part of the experimentation). This practice of single, domestic authorship began to change, slowly at first, but more quickly over time as a number of scientific and social factors began to converge. In terms of US papers published, where just 6% had foreign co-authors in 1980, by 2000, more than 25% were internationally co-authored. The United States remained (in percentage terms) among the least internationalized of the scientifically advanced countries, thanks to the size of its own domestic scientific workforce: it is easier to find a co-author in the United States, so US scientists are less likely to look abroad. At the same time, again in percentage terms, other countries and particularly the EU became much more likely to have researchers co-author at the international level.

Among the converging factors influencing the rush toward global science is the practice of science growing worldwide, thanks to increases in funding, training, and communications of research. Many more countries commit funds to research and development than at any other time in history. As the amount of scientific research grows, so does the publication of scientific data and findings, which are proliferating at exponential rates. New possibilities for publication posed by the Internet have combined with creation of new venues—sources reporting on science that seem to multiply almost daily—increase the possibility to work across international borders and boundaries.

Over the decades of the late 1990s and early into the new millennium, many governments around the world—responding to the perceived significance of scientific research to economic growth—increased R&D spending. (Many developing countries were helped in this endeavor by the World Bank and philanthropic donor organizations.) For example, in 1990, only six countries were responsible for 90% of R&D spending; by 2010, the number of countries responsible for that 90% had grown to 15. The total amount of R&D funding had risen, as had the number of countries making significant contributions to science. According to UNESCO, in the first decade of the new millennium, global spending on R&D has nearly doubled to almost a trillion dollars, accounting for 2% of the global domestic product.

Among those countries increasing spending for research, the developing countries stand out—they have more than doubled their R&D, according to the United Nations. The number of scientific papers published by developing countries has grown—particularly from Brazil, Russia, India, and China (BRIC). As a whole, for all countries, the number of scientific articles registered in journal databases has increased over the 20 years, from 1990 to 2010, from 600,000 to 2 million. As part of this growth, science at the international level—counted as co-authored publications—continues to grow in raw numbers and as a share of all scientific publications.

Ironically, for the scientifically advanced countries, the growth of scientific research around the globe has meant that their citation levels appear to remain flat. This is an anomaly of the data: as international collaboration has increased, so has the share of credit for publication and citation activities. Credit shared between two countries will make the leaders' growth look slower over time because the counting process means that two countries share the credit. The newer members will look like they are growing much more quickly, but it is their *share* that is growing. As a result, the number of papers that list authors from the advanced countries has flatlined, and any rise in attention to their papers has been due to international collaboration.

The scientific journal article as a method of conveying research results has been the standard medium of formal communication for centuries. It remains the standard today. Although it has had several different names or forms, the article as the principal means of communicating results dates from the sixteenth century when Galileo Galilei wrote pamphlets describing the results of his inquiries. Scientific papers were rarely co-authored in earlier times, even when the research behind it had been collaborative. As institutional laboratories created opportunities for science to become a profession, publication records became important for academic careers. This continued to be the common practice until the later part of the twentieth century when more co-authorship becomes apparent.

COUNTING MEMBERSHIP

Since 1980, co-authorship of scientific papers as a share of all papers has risen at a rapid rate. Part of this rise is surely due to an increase in actual on-the-ground collaboration among researchers, but part of the shift is due to greater willingness to list more than a single author on a paper. Cooperation is not new to science, and cases where cooperation had occurred as part of the research process in the past were underreported. So the rise is partly due to a change in authorship practices and partly due to greater scientific cooperation, although the share that would be attributed to either one is not known.

We can assume that there are many levels of collaboration and cooperation in science, of varying duration and intensity. In the broadest sense, researchers are cooperating with one another through acts such as sharing samples, swapping data, or offering standards of measurement. These would be uncountable in number and certainly outnumber the formal articles that get to the level of a scientific publication. Some of the levels of cooperation are so informal that their influence on scientific outcomes is never acknowledged. Other levels of cooperation and collaboration can include citations to another's work, or acknowledgments at the end of an article or in a book to thank another person for help.

More formal cooperation—those activities over and above the unacknowledged assistance but not necessarily rising to the level of collaboration—can include actions such as reviewing a colleague's written paper or co-designing a symposium or a workshop. These projects can result in research advances for both (or multiple) parties in a cooperation. The outcomes may not converge toward a single answer to a specific problem, but they may provide useful inputs to different scientific research activities. These levels of cooperation may be acknowledged in a scientific paper (in the section of an article set aside for acknowledgments) by listing people's names as a way to thank others for help and signal to readers with whom one has consulted. These kinds of acknowledgments can also be counted and linked to one another as indicators of communication, but are less formal than the co-authorship. New methods for counting and measuring collaboration are a goal for those who study science.

While acknowledging that cooperation and collaboration have varying levels of formality, counting articles is commonly used and widely accepted as a measure of scientific output, as we have discussed. A number of scholars have asserted that the submission of manuscripts to reputable peer-reviewed journals remains the crucial outcome of science, representing findings that the authors collectively are willing to claim as notable. The question of whether the databases that catalog these scientific journals and articles represent science as a whole is a discussion for later. These are the pieces of evidence used to trace the global network for now, but this will likely expand as more researchers use blogs, videos, popular press, Twitter and other social media to track scientific output.

NETWORK DYNAMICS BEYOND THE NATION-STATE

To get a good sense of the growth of international collaboration, I took part in an international collaboration of my own—one focused on counting the extent of co-authorships that can be found across countries. Together with colleagues from Europe and Asia, using data based upon records from the scientific abstracting services, we created counts of countries as they appeared by addresses attributed to authors. Files of thousands of addresses representing article co-authorships enabled us to count the number of times an address appears as sole authorship, domestic co-authorship, or international co-authorship. Once the international co-authorship is counted, it is possible to count for which countries are represented and develop networks that represent the co-authorship relationships.

For these data in 2011, the articles listed in the Web of Science produced a list of more than 2 million separate addresses represented in the papers published in that year. The graph (insert) shows the count of the top 20 most frequently represented country names.[1] The United States appears first, by sheer weight of the size of its scientific workforce; this is to be expected. Further, a visual inspection of the findings suggests that the European Commission's policies requiring that at least three EU countries participate in publicly funded research project have had a significant impact on international linkages within Europe—at least in terms of country counts where European collaboration can be seen as significant and robust over time. China also appears as a newcomer at the international level, where in the past China was not a major global-level participant.

These data can be organized so that it is possible to attribute documents to countries to get a count of publication patterns. Given the vastly different sizes of the countries involved (whose outputs are influenced by the size of the workforce, spending, and numbers of publication outlets), the data are normalized to weight the contribution of each country at the global level. (If counts were based on size alone, the United States would dominate the numbers at the country level in every aspect, although China is becoming an increasingly prominent participant at the global level.) Historical patterns also ensure the participation of the European countries in any core group of collaborating countries at both the observed and normalized levels.

Once the country counts are weighted, it is possible to analyze the network of linkages that are apparent among co-authoring countries. The points of interest for the network at the global level are whether the network is favoring only certain countries of the world, whether it is open to new members, and the extent to which knowledge may be able to "flow freely" across the network—indicated by multiple, inclusive links. This can be measured by examining the density of links across the network, and the extent to which the nodes cluster—that is appear in bundles of highly redundant connections.

---

[1] The count is shown using both fractional (each address gets a percentage share) and integer (each country gets a count of one) counting for 2011.

Following that, we examined the extent to which some nodes within the global network are more "between" others, but appearing closer to the center of a network. It can be measured for the network as a whole and for individual nodes as a betweenness measure. This measure can reveal the power dynamics in a network, as in, no set of nations is operating as more "between" or more powerful than others in a translational sense. It can measure the opportunities for knowledge to flow to peripheral nodes.

One feature we searched for is whether the network has a high or low betweenness. Exchange across a network can become distorted by a strong core with high betweenness—where some nodes have greater power to "pull" new members into their cluster—where the nodes act as a go-between for other nodes. In this scenario, power becomes concentrated, and in the global science system, one might expect to see this. Even though the network is self-organizing, it echoes the political system at the next lower level, and therefore, influence can revert to traditional participants, such as those from the larger, scientifically advanced countries, at the expense of the smaller, newer members, although this assertion needs more testing.

Again, the global network has self-organizing features—it is not responding to a design. Most of the linkages are formed by people within one country seeking and sustaining collaborative research ties with people in another country. The links are based upon the needs of their research rather than the interests of an organizing or governing body. The collaborators most often have their own funding before they begin an international collaboration, so clearly the network will reflect, to some extent, the interests of the political layers underlying it, because the funds are allocated along the lines of missions and goals of political units.

The betweenness measure has shown different dynamics at different points in my measurement of it over time, but the 2011 data show a fairly open network that is able to attract new members and sustain global collaboration among the participants. The network appears to have become more open to new members and much more attractive as a locus of research than at any time in history.

The interesting features lie in the dynamics of change, but these can be quite difficult to model, so I have looked at static snapshots of the network over time. I have found that the international collaborative network has a distinct structure that can be studied, just as we have studied the self-organization of the Internet (which is the largest self-organizing network in history). The Internet revealed to scientists many of the underlying features that can be used to understand the global network of scientific communications. The features have fed back in theory that now helps guide expectations of what to look for in large-scale networks.

Theory tells us that the macro-behavior of a network is not the result of the micro-features of the agents—in this case, the intentions of an individual researchers or nations. The individual joining in or dropping out of the network at the global level does not affect the macro-structure. The formation and persistence of structure is notable because it becomes the functional, organized system operating independently of the agent. The macro-structure (rather than public or institutional policy) determines whether the network is open to newcomers, whether it is creating opportunities for sharing information that can advantage even the less powerful or newer members, whether knowledge flows freely, and whether "local search" for new knowledge (connecting to friends-of-friends through small worlds), resources, or new collaborators will be fruitful.

Thus, we can assert that the global network has a regular structure that has regularities of its own—ones that can be discovered and tested to see whether it has properties that continue to serve the goals of those seeking membership.

In viewing the core of the co-occurrence network over the 20-year time period, a divergent outcome can be observed. At the level of co-occurrences, the number of collaborations in the core nearly doubles, suggesting that collaborations among the scientifically advanced countries are increasing. When the data are normalized in terms of cosine relations, a smaller and tighter network of collaborating countries is found at the core of the global system, but the core is operating fluidly, with new nations joining it. In network terms, this suggests that the core group may be forming into a more cohesive cluster, but with features of an open system. There may be several reasons for the formation of a core group, including the possibility that policymakers are shifting their choices to exploit the possibilities offered by the global level. In other words, as actors began to observe globalizing forces during the 1990s, many of them shifted their choices to incorporate a wider view of the system, but with policy discretion as to which countries with which to connect—most likely seeking to collaborate with more elite groups, as preferential attachment theory would suggest.

There are a number of possible reasons for the creation of the global network and its highly interactive group of nations. Although the circumstances were favorable for the global network to emerge as the Internet was created, it is not the inevitable outcome, and the Internet did not create global science. Many aspects of life have not evolved toward global networks, even though the physical ability to do so is there—there are features of scientific knowledge creation that led to this development. The reasons may be important because now the global system is dominating national or local science in unexpected ways.

To be sure, the science system is influenced by many factors, including public policy, funding, location of equipment, location of key people, access to natural resources, and institutional prestige. Any of these factors might have had an influence on the direction of the global network, yet, it appears that the network emerged from the interests and needs of science rather than other factors. In an article I wrote with Loet Leydesdorff, as the network first began forming, we discussed the possibility that the growth of the core network may have been due to the confluence of socio-political factors occurring in the 1990s, including the breakup of the Soviet Union, the reunification of Germany, the formation of the EU, and the development of information networking. However, if the social-political factors were predominantly driving the network growth, one would expect to see a leveling-off of growth trends at the international level during the 2000s as the former Soviet Union countries and East Germany became fully integrated into world science. However, a leveling-off of connections is not what happened: by adding more years of data to our earlier analyses, we see that international collaboration has grown and the network has become denser. While the entry into the world science system by former Communist countries may have played some part, this is not the most significant reason for global network growth.

The development of the network core upholds earlier findings by Albert & Bárabasi (2002) that the development and evolution of a large network is governed by robust self-organizing phenomena (another way to describe emergent properties) that go beyond the particular interests of the participants. It is important to note, however, that the emergent properties serve the overall interests of the participants or they would not be attracted to participate. This may not always be apparent to the participants, for as Powell (2003) has pointed out, at some point in the evolution of a networked structure, all parties to a network can lose some of their ability to dictate their own future: they become increasingly dependent on the activities of others.

Take the example of physics experiments on the Large Hadron Collider at CERN, in Geneva, Switzerland, Europe's center for high energy physics, and an instrument-driven center. In addition to the unique particle accelerators and colliders at CERN, computers reign as the most transformative instruments. According to Rolf Linder,[2] a team-leader at CERN, computing has transformed physics experimentation over the past 20 years. The core team of 10, which he leads, meets every day, face-to-face. The core group continually coordinates with larger groups to design and build equipment, ensure safety, and create requests for equipment time. The core team shares details with a core-plus-one team that is on-site at CERN, and which meets once a month—this larger group of another 10 people chooses and schedules experiments. Beyond this, a core-plus-two team of about 15 more people is dispersed in member-country laboratories. Once a country "joins" an experiment, that country assigns a scientific representative to interact with CERN and begin sharing in the planning for experimentation; the LHC team competes with four other projects for time on the accelerator. The larger team meets as a group once a year. But it is the abundance of data that can be stored and quickly shared over the Internet that has really changed the way they work: "The amount of data has changed the pace at which we can develop new experiments," Lindner said, and thus drives the competition for the resources. "We need computer scientists on every team," he said, "and they process and share data with the relevant team members."

The CERN networks of teams-within-teams and the multiple connections among them provide a relative independence from other parts of the institution. Lindner's teams become a resource to others conducting physics in dispersed locations. CERN may be viewed from one angle as a hierarchy, but Lindner's networks exist within and around it. These features of connections that defy the classic hierarchy also make networks hard to study—they are different from command-and-control organizations, and input-output or cause-effect relationships, which have been intensively studied and are fairly well understood. Some may be less comfortable "managing" networks because organizational behavior textbooks spend very little time on them. Science stands to benefit greatly from understanding complex systems and their networks, if for no other reason than that science operates effectively and efficiently as a network, and has done for four centuries.[3]

[2] Personal interview, July 9, 2013.

[3] This historic role of networking in science was the theme of my book, *The New Invisible College: Science for Development*, Brookings, 2008.

Grappling with complex systems has frustrated many managers and policymakers in part because people have a limited ability to deal with complexity. Thus far, the human tendency has been to stretch existing tools and mechanisms made for command-and-control system of management to try to manage complexity—with limited success. This tendency is rational because people tend to use the tools on hand, but the outcomes may not be as expected. Better we should examine the emerging system we are trying to manage.

INTERNATIONAL COLLABORATIVE PROJECTS

The global network emerged to its current level of organization due to the interests of a number of different stakeholders and for diverse reasons. International collaborative projects take many forms. I've identified four of them.

Megascience

Endeavors that have traditionally been called "megascience projects" have led the way in international science, mainly because the cost of the undertaking was so large, no one nation could afford to do it alone. In addition, these projects, such as the International Space Station, have the added benefit of offering an opportunity for rival nations to cooperate on a peaceful level. The soft diplomacy of megascience projects has sometimes made them more popular with the foreign office than with the scientific sponsors. Such megascience projects are often "top-down" in organization—where governments that are overseeing national interests in global scientific outcomes negotiate their design.

Megascience projects include the Large Hadron Collider, the Human Genome Project (HGP), the Hubble Space Telescope, and ITER (the international nuclear fusion project). National governments traditionally fund much of these projects usually paying into a common fund, although there are projects where each nation contributes equipment to a common project rather than to a common fund. Usually, these projects involve huge pieces of equipment that must be located in one place or in a series of places in a coordinated way. These projects tend to be long term, multidisciplinary, and involving many different teams that may be applying for funds separately.

Megascience Without a Central Organization

Megascience projects should be distinguished from large-scale, globally distributed projects, such as the HGP. HGP is an example of a project that involves international teams and collaboration of many kinds, but each nation funded its own work participants in their own labs, and the research was generally conducted within that specific country. Coordination was done on the Internet and by phone. In fact, during the project, the team leaders coordinated by phone on a daily basis to share findings and shift the direction of research, as needed. Data were shared in large-scale data repositories that were available to different teams working on distributed projects. The teams could span national boundaries as needed, but they formed and dissolved as the research changed over time.

Small-Scale Research Projects

In something of a contrast to megascience projects, global distribution of smaller-scale research has been the sector of global science that has shown the greatest growth, due to the Internet and less-expensive travel. These projects are mostly self-organized, bottom-up projects where researchers choose to collaborate with a scientist from another country (or more than one country) based upon the needs derived from the science itself. The bottom-up feature is facilitated by the needs of the researchers who often report having met at a conference. Projects can be carried out separately or together in shared lab space, and any combination thereof. Many times, researchers will share doctoral students or post-doctoral researchers, who can sometimes travel more easily than scientists with university or business commitments can.

Resource-Based Projects

A fourth type of international collaborations is one motivated by the geographic location of a specific resource. The Antarctic Research Centers is an example of a case in which international collaboration is mandated by the unique resource offered by in a place of interest. Other examples include the NOvA neutrino research project that shares deep earth mine sites, several IPCC global warming research projects on the world's oceans, and the ALMA telescope project that takes advantage of the environment at Atacama, Chile. Many times, researchers from different places meet at these locations and find reasons to continue a collaboration beyond the time when they jointly visited the international resource. Again, these often self-organized, self-funded collaborations have features of bottom-up research.

Within the many different countries that have joined the global network of science, the project profiles are likely to favor the bottom-up research activities. The megascience projects can be too expensive for most developing countries to pay into. Scientific societies will fund researchers from developing countries to participate in megascience projects, but the developing countries are rarely involved at the governmental level. Scientists from developing countries are much more likely to be involved in the globally distributed and geographically tied projects in large part because these researchers have had a chance to meet their compatriots from other countries, find common interests, and subsequently retain a link through social communication tools.

The numbers of global projects in progress at any one time is difficult if not impossible to know. Publications that result from joint projects serve to indicate the extent of international collaboration, but even here, there are questions about how to count collaborations and what is truly represented by the numbers. As scientific research investment grows across the globe, we can expect that the publication of scientific data and findings are proliferating, and as I have noted earlier, this is very much the case.

The counts that have been made of scientific output use databases created to track and monitor science. Several of these databases are used to provide insights into science. Some are maintained by private companies for subscription and sale, and some are maintained by governments. The history of these databases is worth a slight divergence from the storyline.

With the rise of the Internet and electronic data storage has come a proliferation of sources on the Internet that track and monitor science, and that collect scientific journal articles. A growing number of journals (paper and electronic) are available—as we discussed earlier, new journals tend to indicate the rise of a new subfield of science. Traditionally, these journals have been created by scientific societies or academic centers and published on paper. The proliferation of sources has been in the creation of online journals, most of which are open source venues, such as arXiv, and conferences (physical and cyber) and is a sign of the health of science. In 1990, six countries were responsible for 90% of R&D spending; by 2008, according to UNESCO, this number had grown to more than 13 countries making significant contributions to R&D. Since the beginning of the twenty-first century, developing countries have more than doubled their R&D spending. This growth bodes well for scientific capacity in developing countries, and the data on publication trends uphold this observation.

These changes present significant challenges to assessment and peer review processes. Indeed, recent calls for global standards for scientific assessment are well intentioned in theory, but the nature of change in scientific communications makes the idea of global standards a dream that would be difficult to put into practice.

The proliferation of sources and venues reporting on scientific findings has energized many fields of science. The electronic journals, newsletters, and bulletins reporting on science do not always use established standards of peer review and editing, and the underlying practice is nearly impossible to sort out. The variation in quality is complicated by the legacy of the abstracting services created to track journal literature. Most users are familiar with the Science Citation Index Expanded (SCIE), a service of Thomson-Reuters, which is an abstracting service that is called the gold standard in scientific abstracting. However, many users are unaware that SCIE represents only a small percentage of scientific publications: in other words, there is a lot of "unseen science" at all levels, but especially at the global level.

In a recent inquiry into the representation of developing countries in SCIE, I found that SCIE lists only a fraction of all journal titles. This includes those journals from both the developed and the developing countries. Among BRIC, more than 90% of publications are not cataloged and abstracted in that widely used database. A review of SCIE found that in 2010, it listed 495 venues published within BRIC countries, while our survey turned up collectively as many as 15,000 scientific publications being published within the BRICs. This means that the network measures are based upon a small percentage of the most elite scientific publications, and the network among other authors on other publications is unknown.

At first glance, the scale of unseen science appeared to be a vast under-representation of developing countries in global science, but amazingly, the 3% rate is not disproportional when compared with advanced countries: our calculations show that most countries, again, developed and developing, have only about 3% of titles listed in SCIE. In other words, to the extent that citations and impact factors use SCIE-listed publications, more than 90% of science remains unseen, unaccounted for, and omitted from the comparison pool.

In the global network, what does it mean if 90% of research publications are not on the network? The implications of this omission pose serious challenges to anyone seeking to create a global standard for scientific assessment or a reasonably comprehensive understanding of the global network. A 2011 meeting of 50 national research agencies called for transparency in national funding decisions, and by extension, greater harmony in the methods of assessment and evaluation of what constitutes good science. Peer review was touted as the gold standard for determining value. But transparency and peer review assume that the work of scientific researchers will be readily available for scrutiny and that the materials being scrutinized are comparable. This is highly questionable under the current mix of practices.

It is ironic that, on one hand, the information revolution makes it easier to access the results of scientific research while, on the other hand, creating the opportunity for proliferation of output venues. The result is a rapidly growing, open system that is harder to track and monitor than ever before. Global transparency would only be achievable with new protocols for viewing, cataloging, and understanding the swath of activities that represent science in this collaborative era. Currently, the vast majority of scientific publications remain unseen by most potential users. Regional databases beyond SCIE—for example, ones that track national publications in the BRICs—are growing in number and scope. Table 2.2 shows the names and country of origin of the largest of these databases.

For developing countries, the difficulty of joining the global network is compounded by the language barrier. China publishes 6596 scientific journals, of which only handfuls are abstracted in English. This makes it impossible for most non-Chinese scientists to read the work. Similarly, Russia and Brazil each has close to 2000 scientific journals in national languages that are not also indexed in SCIE. India is better represented in English, but unlike the other three countries, its national publications are scattered among a number of databases that are difficult to track down.[4] For scientifically advanced countries, such as the United States, the non-SCIE venues are more readily found, usually through online discipline-based collections, such as the Index Medicus, Chemical Abstracts Service, or the subscription-based Ulrich's Periodical database, but no single

[4] Wagner, C. S., Wong, S. K. "Unseen science? Representation of BRICs in global science." *Scientometrics*, 2012. DOI https://doi.org/10.1007/s111192-011-0481-z.

source provides common access (with Google Scholar coming closer than others).

Science policymakers and academic administrators rely upon indices of impact and recognition to aid them in adjudicating among competing priorities for funding and promotion across a broad swath of science: as non-experts, they must rely upon the judgment of peers to determine quality. This determination has historically relied upon the primary journal literature as that literature communicates basic scientific research findings. Assessments of journal impact and citations to articles are commonly used within and across disciplines and nations to show productivity and create rankings in outlets, such as NSB's *Science & Engineering Indicators* report and UNESCO's *Science Report*.

These rankings draw upon the commercial abstracting and indexing services, such as the SCIE. The design of these services was well suited to paper-based journal publishing that has formal processes for review and editing. However, as new venues for communication proliferate, the abstracting services cannot keep up with a great deal of scientific output. This problem only grows as new Internet-based venues, such as arXiv and Researchgate, grow in popularity. These recently developed communication tools enable new participants to join into scientific communications, which is a good thing. However, they also challenge traditional bibliometric analyses to truly account for the majority of scientific activity. How much of scientific communication and ranking is unseen? By our count, it is quite a lot.

Science policymakers, administrators, and historians of science are interested in knowing the number and form of scientific publications. It is widely recognized that the communication of basic scientific research depends "almost entirely on the primary journal literature" (Braun et al. 2005, p. 95). The journal publication system has been remarkably stable over more than three centuries. Rankings and assessments of journals are commonly used (1) within and across countries to show productivity and create rankings in outlets, such as the *Science and Engineering Indicators* report (NSB 2011); (2) across institutions in reports, such as the *Research-Doctorate Programs in the United States* (NRC 2011); and (3) in individual promotion and tenure decisions.[5]

---

[5] In P&T decision-making, the journal impact factor becomes important. The JIF is calculated. The impact factor is a measure of the frequency with which the average article in a journal has been cited in a particular year. The JCR also lists journals and their impact factors and ranking in the context of their specific field(s).

Many thoughtful scholars have inquired into the extent and scope of scientific publishing. While journals remain a principle mode of communicating research results, over the past 20 years, new modes of publishing and new participants have been changing the system of scientific communication. Scholars have noted the diversification of the modes of scholarly communication and an expansion in the accessibility of research results through open access venues. Many more countries of the world are funding research and development, notably the BRIC countries.

REFERENCES

Albert, R., & Barabási, A. L. (2002). Statistical Mechanics of Complex Networks. *Reviews of Modern Physics, 74*(1), 47.

Braun, T., Glanzel, W., & Schubert, A. (2005). A Hirsch-style index for journals. *The Scientist, 19*(22). Letter, accessed Sept 2018, https://www.the-scientist.com/letter/a-hirsch-type-index-for-journals-48137.

National Research Council. (2011). *Research-Doctorate Programs in the United States*. Washington, DC: National Academies Press.

National Science Board. (2011). *Science & Engineering Indicators*. Washington, DC: US Government Printing Office.

Powell, W. (2003). Neither Market nor Hierarchy. *The Sociology of Organizations: Classic, Contemporary, and Critical Readings, 315*, 104–117.

Wagner, C. S. (2009). *The New Invisible College: Science for Development*. Washington, DC: Brookings Press.

CHAPTER 6

Openness in the Global Network

Dense networks—ones that support innovation through many opportunities to search for new ideas—thrive on open sharing of data, information, and knowledge. Openness and reciprocity are required of participants: this applies to all those participating within the network, as well as those aspiring to join it. The aspirant who wishes to join the network must have something to offer (even if it is just time and expertise), and those actively participating nodes and hubs all follow network rules of reciprocity—give in order to receive. The aspirant who has nothing to share will not be "invited" into the network: in other words, no "free riding" in this system. The node that does not offer resources, seek to exchange value, and engage with others will soon find itself pushed out of the network. Hubs that withhold or distort information, that seek to manipulate others, or that exploit the network for private gain will soon find that participants avoid connecting with them and, thus, pathways go around them. Any parts of the whole network that do not openly engage will become brittle and vulnerable to chaotic and competing forces for access to the useful resources within it and, eventually, will lose those resources to others.

Openness has several meanings in a network context. It is both a set of practices requisite for participants and a feature of network operation. On the latter point, the boundaries of the network itself can be described as reflecting levels of openness. The global network of science includes participants who work on international collaborative projects, but it does not

© The Author(s) 2018
C. S. Wagner, *The Collaborative Era in Science*, Palgrave Advances
in the Economics of Innovation and Technology,
https://doi.org/10.1007/978-3-319-94986-4_6

have boundaries in any sense of the word. Any knowledge system could be called "open" if information, energy, or materials can transfer across its borders. The global network is connected to others through each node, which most likely maintains several memberships in institutions, organizations, and networks. A researcher working on a project with a colleague from a foreign country will also have connections within his or her home institution. Similarly, project-based groups (e.g., research teams) share resources among themselves that are most likely shared within an institution. Data from genomics research projects are shared with home-based researchers as well. Knowledge from local and physical connections flows into and out of the network through the multiple connection of nodes that are in turn connected globally, providing ideas, energy, and materials that assist the network.

A second meaning of openness is one applied to nodes in the network—it is the willingness of individual nodes to share information, energy, or materials with the network. It is a concept similar to exchange or reciprocity in network terms. We can assume that the members of the global network—those identified within the institutions listed on co-authored articles—have shared information, energy, materials, as well as findings, insights, and data with each other. The sharing being conducted is likely complementary, in that each partner is providing information, energy, and materials that are additive to those of others, rather than competitive with them. A project where individuals compete for resources may have some functionality for identifying roles within a project, but eventually the group will define complementary roles for each member. This reduces what is called dynamic transaction costs—those costs of project participation related to persuading or teaching partners that are part of a collaborative group that lacks a hierarchy.

A third meaning of openness in a network is the extent to which groups form into clusters or cliques that are relatively closed off to outside influences. The openness of a group will evolve across the course of a project, from being highly open at the beginning as the group forms and brainstorms ideas, to developing into a norms-based relationship where tasks are differentiated and distributed and resources are committed, to task completion and adjourning. Across the life of a team or project group, the extent of openness will shift toward a closed posture while the team accomplishes its tasks. This is not the opposite of openness; rather, it is the necessary process of communication that is required for group identity and task accomplishment. Nevertheless, a clustered network (one with many tight cliques with few links between them) will exchange less information, energy, and materials. This may be functional for the clique, especially in cases where there is a fear of losing control of information, but it is dysfunctional for the whole and it will hinder the dynamism of the network.

Examples of open sharing of scientific data and results abound on the Internet. Ensemble documents genetic data for widespread access and use. The Public Library of Science (PLoS) offers an influential commons-based, peer-reviewed scientific publication exchange. The Creative Commons offers a choice of six licenses that can be agreed to online, allowing users to toggle on specific protections for their shared information. The Faculty of 1000, ResearchGate, and Academia all offer places to share data, information, and knowledge, as well as discussion space for researchers. ArXiv.org, the online preprint service initiated by Cornell University, has expanded far beyond its original mission to physics, and now incorporates papers from a very wide range of scientific and technology fields. As Diane Crane (1972) found, "openness to external influence plays an essential role in the process of innovation in scientific communities" (p. 99).

A knowledge network such as the global science network shares resources (both informational and physical) for four broad purposes: to enable nodes to search for additional resources, to allow nodes to establish new connections with those that offer value, to diffuse information and enforce norms throughout the connected group, and to truth-test and reject false information. Openness ensures that these functions are performed at optimal efficiency. The members of a knowledge network innovate or enlarge their knowledge space through sharing. Each node engages with other partners to acquire ideas, resources, and complementary skills to undertake tasks. The more the whole network shares information, energy, and materials, the greater the overall value of the network to its members. The offer of this potential benefit brings pressure on network members to openly share know-how. As the function of the whole network transitions into a state of greater abundance, the more valuable is the network to its members. Ideas, resources, and skills flow through the network. Nodes with varying absorptive capacity pick up those goods that they can use. They may give back other goods in exchange. By expanding opportunities of combinations of previously disconnected bits of knowledge, skills, or resources, the whole becomes greater than the sum of the parts. The system benefits from the active exchange and gains knowledge capital at the network level, and the network becomes more attractive to others and more influential.

The open state of the network means that engagement is possible with those outside it. Examples of "citizen science" are growing and in turn contributing to science. As Yochai Benkler (2004) has pointed out, some of the best known of these efforts have significantly advanced science: Folding@home and Genome@home are games of distributed computing created to encourage volunteers to attempt to solve knotty problems. The games solve two problems at once: the need to use distributed computing for processor-hungry problems, and access to creative problem-solvers who have untethered motivations to participate in science.

Resources in an open network are shared to enhance efficiency of search. It is often the "outside" idea that stimulates new thinking. In the past, there have been significant costs associated with identifying, assessing, assimilating, and applying knowledge from elsewhere. Networking provides access to what Eric von Hippel (1994) calls "sticky" knowledge—experiential skills that are the result of learning by doing. New information and communication infrastructures are increasingly supporting these searches for new knowledge and innovative processes. These technologies allow for ideas to be worked on, exchanged, and diffused rapidly with decreased transmission costs. Similarly, enhanced software supports a larger potential range and number of participants in projects than has been practical in the past. Certainly, it is easier to learn about a person's reputation and past research now that the Internet searches often find detailed career and publication records of others. The ability to use the network to search has broken down some of the borders of what were once called "tacit" and "explicit" knowledge by Michael Polanyi and others. Search becomes more fruitful and less costly.

Openness is a non-market exchange, but like the exchange of market-based goods, there are costs and benefits of engaging in network-based exchange at both the bidirectional (node-to-node) and multi-directional (whole network) levels. The "costs" are loss of control over the process of acquiring and sourcing an idea; sometimes a longer time is required to complete a task. This concept is based upon the knowledge-creating network, where concerns about protection of intellectual property are of lesser concern than the access to new knowledge. The idea here is that—far from a "build or buy" strategy where nodes calculate whether to conduct a project on their own—the projects in a knowledge network would simply be impossible to complete without the collaboration. Thus, the network does not provide a substitution effect—the work would not have been done other-wise. The complementarity creates new wealth as "knowledge."

Nodes with plentiful resources and expertise attract partners. They attract people who need resources and who have things to offer (such as time to work on a project). Just as has been found for firms by Walter Powell et al. (2005), individuals gain expertise by linking to attractive partners, which, in turn, enhances reputation for both parties. In studies of networks of firms, R.C. Allen (1983) examined the nineteenth-century historical contexts of industry-shared designs and performance metrics of items such as boilers, blast furnaces, mines, and steam engines. The ability of the designers to build upon each other's work resulted in a steady stream of incremental innovation across a community of firms. Sharing led to stepwise improvements which built upon one another. In these cases, the willingness to share may not have benefitted the individual at a specific moment in time, but the collective result was improvement for the indus-try at large.

Hubs are able to enforce norms such as openness and sharing on less-connected nodes. Unlike diffusion networks (whose measure shows how quickly innovation is taken up in the marketplace with regularities for the roles of hubs (see Raynaud 2010)), networks for scientific exchange have the function of both diffusing ideas and relaying the results of experimen-tation and the creation and application of data as they relate to hypotheses. Thus, simple logistics models cannot provide direct assistance to us in understanding the role of hubs.

Although the network is open in a systems sense, this does not mean it is inclusive. In fact, the dynamics within networks can create exclusionary conditions in several ways. The first is participation in a community of scholars. The output of collaborative research is written for publication in an academic journal. The written material is published in journals that may be disseminated widely, but this does not constitute "openness" in itself. Most journal articles are written for other experts, and the use of language constitutes a passive exclusion—most people will not be able to under-stand it. It is not a "pure public good" (non-rivalrous, non-excludable) in any sense of the term used by economists to describe market-priced goods. While it is very difficult to exclude users from accessing published materi-als, this does not constitute a "commons" readily available to all comers, since the technical nomenclature will deter most readers. The knowledge is available to those with the capacity to absorb it.

The second way in which the network can be exclusionary is the action by others within the network to actively exclude some aspirants. As noted earlier, aspirants must have something to offer to the network in order to be welcomed into a collaborative project. The "goods" being offered can include reputation, which is a powerful "good" to offer to others in a network context. It can also include something as simple as a good scientific question, local data, or skilled labor of trained doctoral students. The knowledge system is not an ever expanding field of relations. In fact, there is a limit to our attention (Simon 1947), a limit to the boundaries of a scientific discipline (Price 1963), and a limit to the number of people with whom we can realistically interact (Dunbar 1992). This means that some who seek access to the network will not find a place within it.

A third way in which the network can be exclusionary may be the action of powerful hubs to refuse to link to certain other nodes. This phenomenon is less likely to be seen in a large network such as global science, because the size of the network affords some levels of discretion to members. However, it is possible to consider cases where political pressures will limit certain types of links; however, we note that links can be seen in the network between Israel and Arab countries, between North Korea and others, and that the United States–China cooperative relationship has grown to be the closest bilateral science relationship in the network. Scientific links have been noted in the past to operate beyond politics, and it appears to be the case in the global network, as well.

Valuable ideas can emerge from anywhere in the global network, thus adding to its value. Smart people are everywhere in the world. Intriguing problems are broadly scattered, as are data. Natural resources such as soil, plants, and ice cores needed to understand phenomena are also dispersed. The network provides the possibilities of connection for those who wish to search for resources needed to advance their own knowledge-creating potential.

The network structure determines the pathways of diffusion of knowledge. Diffusion of knowledge is the process, defined by Everett Rogers (2010), by which an innovative idea, information, data, or energy is communicated through certain channels over time among the members of a connected group. Rogers complained that few studies had been done on how knowledge is diffused through a network. From a structural approach, the constructed network determines the pathways by which knowledge flows. The extent to which nodes are already connected determines the probability that nodes will connect to it. Incentives for resource sharing determine how and why it flows in certain directions. In other words, the typology of the knowledge network determines the diffusion process.

INTELLECTUAL PROPERTY RIGHTS WITHIN THE NETWORK

It is impossible to discuss openness without addressing the question of the protection of intellectual property—particularly in formal methods such as patents—and the role they play in the network. Patenting is the idea that an inventor will be granted exclusive rights to use their invention in exchange for disclosure of a novel idea. In the United States, the right to patent is enshrined in the constitution. Scientific expression is generally non-proprietary. It is well documented that scientists have a number of motives for their work, but remuneration is rarely listed among them. Yet, the role of intellectual property in science has dramatically increased over the past decades. Changes to law allowing government-funded research to be privately held and licensed have created a covetous atmosphere, at least in some institutions. This trend has emerged even though there is weak evidence between patenting and social welfare. Very few public institutions have been able to earn returns from science-related IP.

Economists will refer to scientific knowledge as "commons based" or "peer production" when no one uses exclusive rights to organize effort or capture its value, and when cooperation is achieved through social mechanisms we have been exploring—for example, network participation. The incentives for researchers to network among one another, as we have explored, replace contracts and hierarchies that guide other firms of human exchange. The "peer production" model sees the contributions of some being integrated into a usable whole—a concept entirely consonant with the global network structure.

As we have explored here, network participation and open exchange rely upon indirect rewards (reputation, access to resources) in place of exclusive property and contract—which would stymie the very life of science. Norms within the global network take the place of legal devices that are used to protect intellectual property. The network operates as a guard against fraud and plagiarism by shunning anyone who is shown to employ these practices. As we can see, being shunned from the global network would mean being cut off from the most elite, leading-edge scientific collaborators, effectively ending one's career.

This internal policing of intellectual property within the network does not address the question of the more formal use of legal means of protection. Patent and copyright law is exceedingly complex, and the extent to which it overlaps with science is small indeed. Nevertheless, the question is often raised in policy circles when issues arise related to collaboration and intellectual property. Who will own any (possibly) resulting IP? It is possible to argue, as Anna Mancini (2006) does, that patent and copyright law are obsolete. We have witnessed in a very short time the dismantling of copyright for all intents and purposes within the music industry. The shift to openness in that industry came about as the result of the flood of electrons streaming past paywalls at a torrential rate. A new business model to capture value was needed—and it was invented.

Similarly, we may be in a position relative to science where the idea of patenting a result of research—especially cooperative research—is simply time-consuming, costly, and ultimately foolhardy given the multiple channels through which information can be acquired. Chief technology officers of knowledge-based companies often talk about racing for the market with their products rather than waiting for patent offices to take up to two years to assess the non-obviousness and originality of their products. Within that time period, the market for the product may have completely turned over and another generation of product be on the horizon.

Patent offices within companies defend territory across space that may no longer be contentious. Patent attorneys litigate and companies swap money in ways that do not create wealth. The cost to a small company can be crippling. It may be time to consider alternatives to patent and copyright law that are more effective at encouraging innovation and invention than is the current system. Meanwhile, the global network will continue to penalize those who withhold information—further reducing the incentive to patent the results of scientific research.

It has been common in the past to associate patenting activity with innovation. Patents are counted and studied. Citations within patents to scientific work are traced and noted (See Ahmadpoor and Jones 2017). Patents force scarcity, and laws enforce that artificial scarcity for a period of time. During the time a product is patented, no other company is allowed to make or sell the product or service to the extent it can be enforced. The patent can be licensed to others for limited use.

Knowledge abundance has increased the pressure on the patenting system. The "gift economy"—the willingness of people to give ideas and knowledge away for free in return for recognition rather than financial reward. In the music industry, despite widespread piracy, musicians produce more music now than ever and quality has not suffered. Publishing has endured the same fate, yet more books are being churned out than ever. The cost of producing the next digital copy is nearly zero.

Mark A. Lemley (2016) noted that a similar open access challenge is emerging around 3D printing, synthetic biology and bioprinting, and robotics. Once 3D printers are available as consumer appliances, they will enable at-home or in-office printing of product copies and replacement parts using designs that could be made available over the Internet. In synthetic biology and bioprinting, new life forms including bacteria and viruses are being created. As Doudna and Sternberg (2017) point out, the ability to make new genetic combinations using CRISPR-cas9 is becoming widely known and available. It will get easier to make genetic changes. Lemley reports that synthetic biologists are developing collections of "biobricks"—individual modules that can be put together, shared, and recombined in numerous ways. Just like music and publishing, development and distribution costs will approach zero. In robotics, the long-predicted revolution may actually be on the horizon. Honda predicts that it will sell more robots than cars in 2020. Robotics depend upon software, which can be copied; the value of robots will be not in the hardware but in the programming, which can draw upon a robust open source community for programming.

Parts of these industries will likely resist the openness that comes along with knowledge abundance. However, the music industry changed its business model to take advantage of the new delivery methods. Those companies willing to adapt to the new landscape have done well. Lemley and Neukom suggest that patents will not disappear. It's still simply too expensive to bring a blockbuster movie or drug to market for patents or for copyright to become completely irrelevant. But increasingly high-cost products "will be islands of IP-driven content in a sea of content created without the need for IP," according to Lemley and Neukom.

Knowledge abundance is challenging many parts of the knowledge system as it flows around boundaries established in the age of knowledge scarcity. Many of these boundaries are no longer effective at accomplishing their initial purpose. In the next chapter, we examine boundaries and discuss how to re-imagine them in the age of knowledge abundance and global science.

OPEN ACCESS AND OPEN SCIENCE

"Open" is often used as a term attached to global science, technology, and innovation. A guide to understanding the terms is discussed here, but with the cautionary note that the definitions and contexts change with some frequency. In theory, science is an "open" enterprise in that the results are shared among interested and knowledgeable researchers. Another view of openness has emerged around the idea that scientific knowledge should be shared more widely than it has been in the past, when it was published in journals that were largely available through subscriptions (on paper) or behind "paywalls" on the Internet, where publishers charge for access. This expanded concept of openness is driven in part by the idea that scientific knowledge can and should be available to be "translated" into useful applications, products, and services more quickly and more systematically than was done in the past. (Readers may also be interested in reading about the DORA statement on open access at www.sfdora.org)

Open science is the practice of making publicly accessible scientific outputs and processes, including publications, models, and data sets. Science-Metrix (2018) reports that half of research articles become available in some form (electronically) within 12–18 months of their publication in refereed journals. Moreover, they report that two-thirds of articles published between 2011 and 2014 and having at least one US author can be downloaded for free. The practice of sharing preprints through online services such as arXiv.org is also growing rapidly. No single source of pre- or post-publication documents the availability of scientific research. Open source materials are generally characterized as being "gold"—where full text of articles is available from the publisher—and "green"—where full text is available from other sources than the original publisher (such as a government or university repository).

The Budapest Open Access Initiative (BOAI) of 2015 makes an aspirational statement about open access to the research results. The BOAI definition of open access is:

> By "open access" to this [research] literature, we mean its free availability on the public internet, permitting any users to read, download, copy, distribute, print, search, or link to the full texts of these articles, crawl them for indexing, pass them as data to software, or use them for any other lawful purpose, without financial, legal, or technical barriers other than those inseparable from gaining access to the internet itself. The only constraint on reproduction and distribution and the only role for copyright in this domain should be to give authors control over the integrity of their work and the right to be properly acknowledged and cited.[1]

[1] http://www.budapestopenaccessinitiative.org/boai15-1

The BOAI statement is controversial in part because some national copyright laws restrict the kind of free and open sharing envisioned in the vision statement. Access and use are being challenged, however. Some research groups, such as those within the field of Artificial Intelligence (AI), are calling for journals to be created online that are both open access and "zero-cost" to the authors. AI and other fields are also discussing a process of "open review" where colleagues can access pre-prints online and then openly post comments and feedback.

Open access policy is institutional support, often by government, university, or non-profit entities, for unrestricted access to research and underlying data as part of the mission to enhance public benefit and translation to application. Policies such as these are not new; many governments have encouraged open sharing of publicly funded research, but the avenues for sharing results have often made it difficult to make information available.

Other uses of the term "open" add to some confusion to discussions about access to information. Adding the modifier "open" does not necessarily mean that information is freely available or that collaboration is determined on a voluntary basis. For example, "open innovation" is a term applied to the concept that inflows and outflows of knowledge, beyond the walls of the laboratory, firm, or team, can accelerate innovation and enhance consumer and user satisfaction with final products. It does not mean the information is freely shared. Similarly, "open data" describes publicly available and accessible data that can be universally and readily found, used, and redistributed free of charge in ways that enhance veracity and research efficiency; however, very few truly open data sources exist. Open repositories are online stores of scholarly research results that anyone can inspect and comment upon, usually in specific scientific disciplines and moderated by experts.

References

Ahmadpoor, M., & Jones, B. F. (2017). The Dual Frontier: Patented Inventions and Prior Scientific Advance. *Science, 357*(6351), 583–587.

Allen, R. C. (1983). Collective Invention. *Journal of Economic Behavior & Organization, 4*(1), 1–24.

Benkler, Y. (2004). Sharing Nicely: On Shareable Goods and the Emergence of Sharing as a Modality of Economic Production. *Yale Law Journal, 114,* 273.

Crane, D. (1972). *Invisible Colleges; Diffusion of Knowledge in Scientific Communities.* Chicago: University of Chicago Press.

de Solla Price, D. J. (1963). *Little Science, Big Science.* New York: Columbia University Press.

Doudna, J. A., & Sternberg, S. H. (2017). *A Crack in Creation: Gene Editing and the Unthinkable Power to Control Evolution.* New York: Houghton Mifflin Harcourt Publishing Company.

Dunbar, R. I. (1992). Neocortex Size as a Constraint on Group Size in Primates. *Journal of Human Evolution, 22*(6), 469–493.

Lemley, M. A. (2016). The Surprising Resilience of the Patent System. *Texas Law Review, 95,* 1.

Mancini, A. (2006). *International Patent Law Is Obsolete.* New York: Buenos Books America LLC.

Powell, W. W., White, D. R., Owen-Smith, J., & Koput, K. W. (2005). Network Dynamics and Field Evolution: The Growth of Interorganizational Collaboration in the Life Sciences 1. *World, 110*(4), 1132–1205.

Raynaud, D. (2010). Why Do Diffusion Data Not Fit the Logistic Model? A Note on Network Discreteness, Heterogeneity and Anisotropy. In *From Sociology to Computing in Social Networks* (pp. 215–230). Vienna: Springer.

Rogers, E. M. (2010). *Diffusion of Innovations.* New York: Simon and Schuster.

Science-Metrix. (2018). Analytical Support for Bibliometrics Indicators: Open Access Availability of Scientific Publications. http://www.science-metrix.com/ sites/default/files/science-metrix/publications/science-metrix_open_access_ availability_scientific_publications_report.pdf. Accessed July 2018.

Simon, H. (1947). *Administrative Behavior.* New York: The Macmillan Company.

Von Hippel, E. (1994). "Sticky Information" and the Locus of Problem Solving: Implications for Innovation. *Management Science, 40*(4), 429–439.

CHAPTER 7

Nations Within the Global Network

To the extent that they have financial resources to commit to it, nations nurture, fund, and use science. To a very great extent, it has been national treasuries and public funds that have built science to the scope and influence it enjoys in the twenty-first century. Even as we see the global network growing in size, power, and influence, nations still arbitrate and fund science for the public good, and they can be expected to continue to do so. How they account for these investments will, by necessity, change as the system changes.

The global network is made up of people working in a place, acting within national political borders. Mainly, they are funded by public coffers. The global network of science can be thought of within the complex adaptive system as a "metapopulation," where we might liken individuals to particles that reside within connected hubs. Individuals can migrate along the links between the hubs with relative ease due to transportation links, the Internet, and rapidly available knowledge. But the particles will retain a national character as they venture forth.

Policymakers care about the locations of national players as well as the outcomes of research. The history of the nation and the history of modern science have been intertwined around funding, prestige, and wealth. Western historians viewed the political form of the nation as coming into ascendency in European history with the Treaty of Westphalia in 1648, around the same time as the first scientific societies were forming in Italy

© The Author(s) 2018
C. S. Wagner, *The Collaborative Era in Science*, Palgrave Advances
in the Economics of Innovation and Technology,
https://doi.org/10.1007/978-3-319-94986-4_7

(the Florentine Accademia del Cimento was founded in 1651). National history has been closely intertwined with scientific prowess since that time. Rankings rose up to compare one nation to another in terms of scientific output and excellence.

Certainly, we know that China operated as a political entity for centuries before the seventeenth century (CE)—and Chinese history also claims inventions like gunpowder and paper—but China did not systemically support science as we think of it today—public funds tied paid for by national treasuries. The Chinese system operated in some isolation from the West for several centuries, and many developments occurred in common, in isolation from one another. China's long period of scientific quiescence certainly kept it from advancing for many years—a factor now being turned around with significant investments in R&D and a rise in output.

Carol Harrison and Ann Johnson (2009) traced the historical links between science and national identify, suggesting that the Western twentieth-century nation-state incorporated the belief "...that scientific institutions would create a scientific public, which in turn would effect necessary future transformations such as the full realization of human rational potential and the creation of a middle class" (p. 3). Such was the idealism that intertwined the political and the scientific national identity, bolstered by a belief that a rational human populous would build the orderly state, supportive of a technocratic approach to governance.

The twentieth century showed that human nature—with tenacious arational features—is less malleable than was proposed by this vision or by the concepts espoused by Enlightenment or socialist philosophers. This is discussed in detail by Francis Fukuyama in *The Great Disruption* (1999), where he asserts that "some norms are grounded firmly in biology" (p. 350), meaning that some behavior is built into human nature and therefore cannot be considered to be completely "socially constructed." Religion is especially tenacious, and in the late twentieth century appears once again in formerly Communist nations. In the United States we find arational rejection of demonstrated scientific evidence about climate change. Magical thinking has not been obviated by science, despite Harrison & Johnson's views and those of others. Neither has spirituality diminished as a human need or urge. These parts of human nature appear to be built into us in ways we only dimly recognize—but part of our basic desires include one for knowing about the natural world, as much as it is to manipulate it with mechanics and techniques.

Scientific thought has not conquered the globe in terms of political/ social approaches to governance. But science has been shown to be excellent at solving problems and creating social goods. This part of science and its R&D efforts are widely accepted across societies. This has enhanced the attractiveness of science within the political system, even within the poorest countries, where the idea of investing in elite sciences—those ventures at the apex of human achievement—is very far from their minds. Thus, we see that science is "globalizing" in terms of investment and worldview at the same time that many other social and political functions may not be widely accepted—such as bureaucratic governance or even voting. The question of whether science operates best within a liberal political order remains to be fully tested. Just as science does not obviate spiritual/arational human nature, the global system does not obliterate the national system. It is possible to observe that tribes and families still operate as subnetworks, even as states and nations arose as more ordered social structures (Fukuyama 1999). Subnetworks are retained, even when state structures deteriorate. Scientific connections can be seen as one of these subnetworks that persist even as states change form or disintegrate. (Some states in Africa, such as Cote d'Ivoire, had robust scientific activity, but when state functions break down through war, the scientific community emigrates to more politically stable places.)

We can say that varying levels of social and political organization continue to operate within and across other forms, such as with the rise of the EU—where new forms of political coordination are added. Feedback and feed-forward loops are then created among the added varying levels. Each new, emergent level might be interpreted as gaining greater power and influence over the levels below it, although new levels are less stable than lower levels, as we see when transnational systems are added: these tend to be much less stable and sustainable over time. Without political constituency, it is difficult to maintain them.

Writing in the early 1960s and describing the science system of his time, Derek de Solla Price details scientific activities using national borders as the boundary; he only briefly mentions international links. The global system of science simply did not exist in Price's time, nor did he offer a vision of science beyond the nation-state. Science took place within institutions solidly grounded within nations, and nations produced science. The outputs supported national goals, and achievements served national prestige. Scientific investments and national wealth became closely correlated and the links were studied by economists. Concepts such as the "national innovation system" emerged to instantiate the connection for policy purposes. It appeared that science was the affordable indulgence of wealthy states. The ties between science and the state established the practice of collecting data and representing science as a national endeavor which still occurs within these categories through institutions such as the OECD.

RAPID RISE IN SCIENTIFIC INVESTMENT

The past 30 years since 1980 has witnessed a stunning metamorphosis for science and technology as a class of activities—once considered marginal to economic growth, S&T has moved squarely into a central tenet of economic growth theories, right up there with "land, labor, and capital." This has occurred as economic growth has shifted toward a "knowledge economy" or knowledge-intensive growth. Land, labor, and capital are no longer viewed as the sole contributors to economic. The correlations between wealth, science, and growth have served as justification for developing countries and intergovernmental organizations to invest in science.

For many reasons—with investment in science and technology being one—since the mid-1990s, a number of countries have experienced rapid economic growth. In the mid-2000s, developing countries whose economies were once tied to the growth patterns of developed countries broke free of that pattern and began growing faster than the developed countries. With the notable slow-down following the 2008 global banking crisis (triggered by the subprime mortgage crisis in the United States), the pattern of growth has still been overall positive, with many developing countries showing strong economic growth in the 2010s, including in knowledge-based sectors.

The twentieth-century scientific powerhouses of Great Britain, Japan, Germany, and France still command strong positions in the any assessments of national and global science, contributing significant funds to S&T, producing high-quality publications, and attracting students and mature researchers to their world-class universities and research institutes. These five countries alone are responsible for 59% of all spending on science globally in the 2010s. These nations enjoy what some call an "inherited advantage" of established scientific systems.

Growth in science is also evident in the expansion of tertiary education and budgets allocated to R&D by relative newcomers to science and technology. Large emerging developing countries such as Brazil, Russia, China, India, Mexico, and South Africa all spent more on R&D in real terms and as a percentage of GDP than in the past, with sharp upticks beginning in the 1990s. Since the beginning of the twenty-first century, global spending on R&D has nearly doubled worldwide, according to

UNESCO, with much of the growth coming from developing countries. As we have seen, the number of indexed scientific publications has grown by a third (based upon Scopus counts), and the number of trained researchers continues to rise. North America, Japan, Europe, and Australasia have all witnessed growth in R&D spending and scientific output, with each part of the world increasing spending by around one-third in the 2000s. In the same period, developing countries, and especially the emerging economies of BRIC countries (Brazil, Russia, India, and China), more than doubled their expenditure on R&D (UNESCO), increasing their combined contribution to world R&D spending by 7 percentage points from 17% to 24%, and contributing 20% of the world's authorship of research papers, again in Scopus data, over the decade of the 2000s.

Total and relative positions of nations in science are measured by a common standard to facilitate comparisons. The databases that underpin comparative studies in output collect a specific subset of all scientific publications and citations. These datasets, as useful as they are, in no way represent all the published output of scientific R&D, as we discussed above. No single source provides access to all of the world's science. But the existing services comprise (and by doing so, reinforce) the most elite science. As scientific practice and outputs have grown, the databases have grown too, with new journals having been added to the collection. This means that comparisons over time are only partly dependable. Indeed, it has been my own experience, in visits to African universities, that quite good scientific research was being conducted, but it remains "unseen" within the journals that publish and the databases that collect world science.

That said, the databases provide the materials we have to work with when examining the positions of nations in the global network. The databases show that relative positions of nations compared to one another have experienced significant shifts over the past 30 years. For example, between 1996 and 2008, the share of the world's articles with an author from the United States dropped to one-fifth of the world's authorship, Japan dropped 22% and Russia dropped 24%—not necessarily in terms of productivity but in terms of shares of total world output. Great Britain, Germany, and France also dropped in relative terms in numbers of articles, but not in terms of impact of these articles on world science, at least as counted by citations to these works made by others.

Increased participation in science is particularly true for China, which has increased its scientific output to the extent that it is now the second highest producer of research journal articles in the world. India has replaced Russia in the top ten, climbing from a lower status in the mid-1990s to the top ten in the 2000s. Farther down the list, South Korea, Brazil, Turkey; Asian nations such as Singapore, Thailand, and Malaysia; and European nations such as Austria, Greece, and Portugal have all improved their standings in the global lists of countries with good-quality scientific results. A cynical analyst might say that these countries have learned the rules of the game in terms of gaining international attention to their scientific output. Rankings and positions are partly engineered through deliberate actions, but the return to increased status is high. It can be seen in the ability to attract students and researchers, international funds, foreign investments, and international conferences, all contributing to a virtuous cycle of knowledge-intensive growth.

With country-level data in hand, it is possible to examine regional trends in global science over time to view whether large shifts occur in scientific output and knowledge absorptive capacity over geographic spaces. Slavo Radosevic and Esin Yoruk (2014) recently found that science in Eastern European countries declined in production and impact during the 1990s, although my data show these countries greatly increasing their *participation* in the global network during this same time period. Data used in both of these measures are difficult to decouple from the underlying political changes—some of these countries emerge from the Soviet Union in the 1990s, and therefore cannot be compared historically because they were not counted separately in the 1980s. Although the publications and citations data show them as "dropping" (OECD 2014; Radosevic and Yoruk 2014) in share, the network data show them becoming more integrated into the global sciences; one can infer that they are therefore more likely to be working at the leading edges of inquiry, certainly more so than they were while isolated from others behind the artificial constraints imposed on science by the Soviet Union.

When examining the publications emerging from these countries, Eastern European countries (minus Russia and Ukraine) saw their share of articles and impacts grow in the 2000s. Their percentage of world publications rose to close to 5% during this time, and these articles attracted more citations relative to other countries than had been the case in the past. The network analysis shows a more optimistic picture of science in these countries than does the publication data. Poland, for example, moves up in terms of betweenness centrality (a measure of their influence in the network), from 34th place in 2008 to 27th place in 2013, which suggests that this nation has become more integrated into global science over that period. Eric Archambault (2010) argues that, of the Eastern European countries, only Lithuania and Estonia have truly thrived in the post-Soviet era in science: both countries saw their scientific productivity rise significantly in the 2000s compared to their Baltic neighbors. They are not highly integrated into the global system, however.

For Russia and Ukraine, the publication data and network analysis agree that this science and technology system experiences a long slow decline in influence from the 1990s until the present. In the 1990s, Russia and Ukraine produced by one count more than 7% of world publications, but this dropped to less than 4% in the 2000s (Radosevic and Yoruk, using Web of Science), by another count from 1% of world share to 1.6% share (Kumar and Asheulove 2011, using Scopus). Russia's ranking in the SCImago Journal and Country Rank also dropped through the 2000s decade from 9th in the world in 2000 to 16th in the world in 2012 (SCImago 2014). Russia drops in betweenness centrality measures overall in the global network, from appearing in the top 20 countries in 2008, dropping to number 36 in 2013. Russian and Ukrainian authors do not have papers among the most highly cited papers in the world (Leydesdorff et al. 2014). No doubt, there are many factors that contribute to this drop of influence and impact of Russia and Ukraine in global science, including perhaps the emigration of many talented scientists and engineers, reduced financial investment in science and technology, as well as other non-technical factors.

The decline of Russia is not uniform across fields of science. Russian world share of publications in mathematics, chemical engineering, agriculture and biological sciences, and medicine grew from 2001 to 2009, but productivity levels do not correlate directly with positions in the network: within the global network, Russia can be seen as holding greater centrality than some other countries in astrophysics, decision sciences, and chemistry. But in the "newer" sciences of biochemistry, computer science, pharmo-toxicology, and neurosciences, Russia has a poorer showing in terms of betweenness relative to other countries. In several key sciences— for example, business and chemical engineering—Russia does not appear to have any centrality in the global network, meaning it is not attracting international partners. The diminishing positions in the network most likely mean that new ideas and techniques are not reaching Russian and Ukrainian scientists, at least not at the rate they reach others, even though these countries remain comparatively "rich" in scientists and engineers.

The EU—both as a single entity and by member countries—has grown to be a scientific powerhouse since the early 1980s, when research cooperation began as a social experiment, one of creating a common European economic market and research area. Europe's scientific productivity shows steady upward growth from the 1980s through the 2000s (Archambault 2010). The early EEC—European Economic Community (as the EU was called in the 1980s)—targeted financial and structural assistance to technology-based sectors to improve competitiveness relative to the United States and Japan. Initial programs such as BRITE and EUREKA did little to improve leading-edge technology in Europe, but they had the effect of drawing together, not the best researchers (who remained on the periphery of these collaborations), but the very good ones. These researchers and their respective units benefited greatly by the association (even if artificially imposed) within collaborative research units, and cooperation had the effect of "bringing up the rear"—improving the contributions of lagging partners. Accordingly, the European Commission continued to encourage and grow cooperation, and eventually, policy mandated that any research projects receiving funding from the European Commission should include participants from at least three member countries. (In the early days, the United States was not welcome in the cooperative projects. This changed with increasing globalization of business. Multinational companies confounded efforts to distinguish a United States business from a European one and increased calls for broader research partnerships.)

As they were admitted to what became the EU, smaller, newer members were afforded special consideration to help grow their S&T capacity. Catch-up countries like Greece, Portugal, and Romania thrived under the cooperative regime. And while it is difficult to ascribe causality to these EU policies, the catch-up countries do improve more rapidly as members of the EU toward closing the gap with the scientific powerhouses of Germany, France, the Netherlands, and Great Britain.

Perhaps even more intriguing for our story is that the scientifically advanced countries of Europe greatly improved their positions relative to the United States over the 30 years that Europe has been encouraging knowledge integration and a common research agenda. Research I conducted with Loet Bornmann (Leydesdorff et al., 2014) shows that several European nations have increased their overall citation shares relative to the United States, and that six European countries (including Switzerland, which is not part of the EU) have overtaken the United States in terms of relative citation rates.

The dynamism provided by networking, measured in terms of co-authorship relations among EU member states, is shown to have produced strong regional integration at that level. The EU countries improved their positions within global percentages of publications from 31% in the 1980s to 35% in the 2000s. And although the individual EU countries see a drop in percentage shares of world publications, the strongest EU member states have further strengthened their positions in terms of citation counts—in other words, more people around the world are reading and citing their work. The same increased strength can be seen in the network data. Among the papers receiving the most citations, these papers are the ones that are highly likely to be internationally co-authored.

At the level of disciplines of science, different European countries reveal different strengths and varying levels of power and influence in the network. When the network is disaggregated by discipline, Great Britain and France are unfailingly in the top five, most central, or between, positions in the network, in every field of science. Great Britain usually appears as the second most "between" country behind the United States. The positions of Great Britain and France as among the most centralized countries in the global network give those nations a great deal of influence over the flow of knowledge. Leading-edge knowledge is readily accessible to their scientific centers, even when they are not conducting the research within their own borders. For France, which is often portrayed as retaining a strong national character to its scientific enterprise, the position of that nation as one of the top three "between" nations suggests that its science system is actually highly international, if not always in practice, certainly in terms of ability to connect. While the integration of the EU can be seen as a factor contributing to the strength of specific countries, the strongest European nations are also the ones most highly connected to the global network. Perhaps both dynamics were contributing to the improved position of the EU over time. The line of causality is not clear, however.

Australia is rarely listed among the top nations in science, and has relatively low investment in R&D (1.6%) compared to its GDP. Reviews of countries with strong science systems often omit it, perhaps because of the smaller size of its scientific community and its geographical isolation. In global comparisons, Australia is often lumped into a class of countries called "Oceania" where its percentages are dragged down in rank by the very small and scientifically backward countries like the Cook Islands.[1] Nevertheless, examining the global network for powerful and influential players shows that Australia is highly centralized within the network, working across and thereby linking many regions of the world. At the disciplinary level, Australia can be found among the top five most "between" countries in 2013 in agricultural biology, biochemistry, business, chemical engineering, decision science, health, medicine, neurobiology, and social

---

[1] The author spent part of her honeymoon in the Cook Islands, which are very beautiful. No scientific labs were visible, however.

sciences. The network position analysis suggests that Australia is punching above its weight at the global level, perhaps finding niches by positioning in a way to take advantage of new developments, even if its own labs are fast followers rather than leaders.

The Asia-Pacific region has shown the most growth in scientific funding, output, and quality of any region in the world over the past 30 years. Starting from a very low base of nearly no participation in science, Asia's contribution has grown by 155%, according to Archambault (2010). South Korea and Taiwan are now ranked among leading world nations in some areas of science. The region as a whole has increased spending in science faster than any other part of the world. The share of world papers ascribed to Asia-Pacific countries has risen from about 12% in 1980 to close to 35% in 2010. Over the past two decades, South Korean science has shown the fastest growth rate of any country except Iran (Archambault 2010). Within Asia-Pacific, strengths in science are seen mainly in those applied sciences linked to industrial development, such as chemistry, physics, and engineering. This is changing over time, however, as the nature of industry changes toward the knowledge-intensive sectors and as the science sectors mature and expand into more fields.

In the 1980s and 1990s, the Asia-Pacific story was dominated by the rise of the Asian Tigers as they were called. Of these countries, only South Korea can really be said to have become fully integrated into the global science system, ranked by the Institute of Management Development (IMD) as number three in the world in scientific competitiveness in 2009 (IMD 2009). In terms of numbers of publications (regardless of international links), South Korea shows strengths in engineering, physics, and clinical medicine. Korea can be seen in the global networks as participating in most areas of R&D, always more central than Russia, for example. In materials science and in engineering, in particular, South Korea is a highly "between" research partner. However, compared to its positions in publications and citations, one might expect Korea to appear higher in network betweenness—it is not highly integrated into the global network, in contrast to Australia, for example. But greater integration may come with time.

Malaysia has increased its national commitment to science and technology over time, and as this has occurred, it is possible to see Malaysia improving its position within global science. At the network level, Malaysia performs better than any other south Asian country—it appears in the top 20 most "between" countries in the network in 10 of the 26 disciplines studies. Malaysia's top performing fields were business and chemical engineering. It also performs well in physics/astrophysics, health, and computer science.

The Asian Tigers were a breakthrough phenomenon in terms of rapid development, but in terms of sheer numbers, their story can no longer compare to the Asian superpower, China. China's scientific production has increased at 4.4 times faster than the world growth rate since 1980, according to Science-Metrix. China's rise as a national science system and in the global network is truly unprecedented in the history of science. The story is unique; China will be discussed in greater detail in a separate section.

The Middle East region has several countries that participate in the global network, particularly Israel, Turkey, Iran, and Egypt. These four countries contribute more than 90% of the scientific output from the countries in the Middle East. During the 1980s, Israel was the only Middle Eastern country with an active scientific system publishing about 5000 articles per year. That rate has remained fairly stable over time. By the end of the 2000s, Middle Eastern countries were publishing close to 45,000 articles, with the growth accounted for by Turkey and Iran. The interesting story here is the one of Iran, which, over the past 30 years, has had the fastest growth of any country in the world in terms of scientific output, besting even China. Science-Metrix (2009) shows that Iran "mobilized its scientists towards the development of nuclear technologies" demonstrated by the fact that "growth in many of the relevant specialties was several times faster than the world average." Significant growth can also be seen in other areas of science, but not at the same levels. Iran is a relatively weak participant in global science (perhaps because its scientists are not welcomed into certain collaborations), but it appears in the global networks in agricultural biology, chemistry, earth sciences, and veterinary science.

In terms of country pairs at the global level, the links between the United States and Great Britain dominate any other pairing in terms of numbers of co-authorship events (2354 in 2008, and 4089 in 2013) and in terms of citations garnered to the work. Co-authorships including the United States and GBR are the most highly cited of any country pair in the world. This pairing is followed closely by the pairing of the United States and Germany, and by Great Britain and Germany.

The United States outdoes all other countries in terms of numbers of partners. In 2013, United States scientists published articles with co-authors from 200 countries. By contrast, Great Britain published articles with 62 countries. Collaboration with the United States also brings with it the added advantage of citation strength. China's collaboration with the United States brings it into the global network, and the highest cited articles for China in 2013 were those published with US coauthors. Despite the fact that many more countries are participating in global science, citations to collaborations heavily favor the United States and European countries. In terms of sheer numbers of co-authorships and citations, Japan is number 27 on the list, and like China, Japan depends upon collaborations with the United States to gain visibility in terms of citations.

OPEN COUNTRIES HAVE GREATER SCIENTIFIC IMPACT

To further examine the positions of nations and the benefits to them of participating in international collaboration, a team of us[2] examined citation data, spending data, and we developed a measure of the "openness" of national science systems. The data were drawn from the OECD and Elsevier's Scopus databases. We had interesting findings about the relationship of countries to one another.

As we discussed earlier, international coauthorships can account for as much as 60% of articles for some small countries where they have a limited national capacity to conduct science. In the cases of countries where interdependence has grown, international collaboration is needed for advancement because costs and human resources needed for top science exceeds the capability of the country to pay for it. Large-scale scientific projects are only the most obvious; there are many small-scale science activities that require collaboration, too. However, no official statistical measure by country accounts for international collaboration, nor do any economic measures exist of its contribution to growth. We wanted to test whether collaboration is actually benefiting countries that participated in it.

Using measures of the extent to which researchers are "mobile" (people move from within a nation, leave a nation to conduct research elsewhere, or return to their home countries), co-authorships and citations, we suggested a measure for the impact of international collaboration in science. To get a sense of the extent to which authors in a particular country were co-authoring with people from other countries, we used what is called a fractionalized field-weighted citation to analyze these citation impacts in relationship to international co-publications and researcher mobility. We followed the recommendation of Moed and Halevi (2015), who suggested that in cases of complex accountability such as international

---

[2] The team included Koen Jonkers of the European Commission, Jeroen Baas of Elsevier, and Travis Whetsell, Florida International University. We published this work as Wagner, C. S., Whetsell, T., Baas, J., & Jonkers, K. (2018). Openness and impact of leading scientific countries. Frontiers in Research Metrics and Analytics, 3, 10.

collaboration, a multidimensional approach should be used. Accordingly, we combined data sets using citation indexes and OECD national statistics to propose a measure of the benefits of international collaboration. We complement Taylor's (2016) approach to measurement—where he focused on broader economic links—but we limited our measure to publicly funded scientific ties, whether formed through international collaboration or through international mobility.

The project started by gathering all articles indexed in Scopus (an Elsevier product) for 2013, with specific calculations of the fractional number of publications, where, in the case of internationally co-authored papers, each country with an address in the paper gets a proportional share of authorship. The fractional number of international papers was used to calculate the percentage of all papers that are internationally co-authored, by country. A second data set included the fractional FWCI for each country, with citations for five years. The FWCI refers to "the ratio of citations received relative to the expected world average for the subject field, publication type and publication year." For example, a score of 1.50 means the publication receives 50% more citations than the world average, while a score of 0.50 means it receives 50% less than the world average when calculating values for individual articles, as it is an article-level metric. For this study, the FWCI values for countries have been aggregated proportionally, generating a fractional FWCI value by assigning a weight to the article FWCI values according to the frequency with which a country appears in the authors' addresses on the paper. For example, an article with three countries each contributing an equal number of authors on the paper would weigh as one-third in the calculation of the weighted average FWCI for each country.

MOBILITY SCORES AND OPENNESS

The OECD reported data from each of its member and observer countries about their research workforce, showing the number of new inflows, returnees, outflows, and percentage of "immobile" researchers. We used data for 2013 on the percentage of mobile researchers, which included new inflows, returnees, and outflows. OECD identified four different groups of mobility patterns. The "inflows" or immigrant scientists refer to the share of the authors who started publishing with an affiliation containing the country under study while initially using a different country as their institutional address. "Outflows," or emigrant scientists, refer to the share of researchers that started publishing with the country under study as their institutional address followed by publications indicating (an) other(s) country. "Returnees" refer to the share of the authors that first published from the country under study, followed by publications from a different country, and finally publishing again with the country under study in their institutional address. "Mobile" refers to all those researchers who have not remained in the same country during the observation period. The mobility analysis by OECD using Scopus data is based on the full career path since 1996 of all authors in Scopus with more than one publication.

An openness index was developed using the mobility data and the percentage of international co-authored articles based upon fractional counting—again, an allocation of counts based upon country names appearing in the author line of articles. The indicators of engagement such as percentages of international collaborations and mobility were found to be highly correlated with each other. As a result, we calculated a Principal Component (PCN) index of the four measures to create a single measure we called "openness" to indicate the extent of international engagement. Here, we see that some countries are more "open" to international engagement than others shown in Table 7.1.

GOVERNMENT SPENDING

We added data to tie research output to government spending. We used OECD data on Government Budget Allocations on R&D (GBARD) by country for 2011. (In a few cases, the data are derived from Eurostat or national sources.) GBARD is generally about 30% of total national spending on R&D (GERD). The justification for using GBARD is to limit the analysis to government spending, as opposed to including private sector spending. We did this because government R&D spending is more likely than all spending to result in scientific publications. Use of GBARD reduces the chances that industrial spending on R&D would be counted, although there may be some number of articles that are funded by industrial R&D funds. Table 7.2 shows government spending in purchasing power parity for 2011 in US million dollars.

There is a close relationship between government spending and output. In other words, the more money spent on scientific research, the more articles are written and published by researchers. This is the case whether we use the government R&D spending, or use per capita spending—the latter measure helps to normalize the size of countries in comparison with one another, but findings are largely the same.

Table 7.1   Countries from more to less open to international engagement

Switzerland
Singapore
Ireland
Austria
United Kingdom
Sweden
Belgium
Australia
Canada
Portugal
France
Israel
Denmark
The Netherlands
Germany
Norway
Hungary
Mexico
Finland
Spain
Slovenia
Greece
Estonia
United States
Italy
Czech Republic
Russian Federation
Republic of Korea
Brazil
Latvia
Poland
Lithuania
Japan
China
Turkey
Luxembourg

The relationship between output and impact is only weakly correlated. In other words, more articles published do not equal more citations. This relationship is seen most notably in the scatterplot by the examining the positions of the higher impact/high openness countries, which are demographically smaller nations including Singapore, the United Kingdom, the Netherlands, Switzerland, Sweden, and Denmark (Fig. 7.1). These countries all scored highly on the measures of openness *and* impact. It may be

Table 7.2   Government Budget Allocations for Research & Development, 2011

Country	GBARD 2011 $MPPP
United States	144,379
Japan	34,105
China	33,268
Germany	30,103
France	19,984
Russian Federation	18,097
Republic of Korea	17,424
United Kingdom	12,902
Italy	12,075
Brazil	11,200
Spain	10,155
Canada	7737
Singapore	7600
The Netherlands	5951
Mexico	5400
Australia	4748
Turkey	4581
Switzerland	3611
Sweden	3276
Austria	2921
Belgium	2880
Portugal	2814
Poland	2688
Denmark	2483
Norway	2474
Finland	2307
Czech Republic	1936
Israel	1480
Greece	909
Hungary	666
Slovenia	352
Luxembourg	279
Estonia	246
Ireland	123
Lithuania	100
Latvia	60

Source: OECD

that, in order to conduct world-class research, smaller countries must cooperate—since funding research across the board is expensive.

The same cannot be said for the attention paid to all published work: as measured by citations, attention (impact) is only weakly related to government spending. Nor is it related to the size of the research workforce.

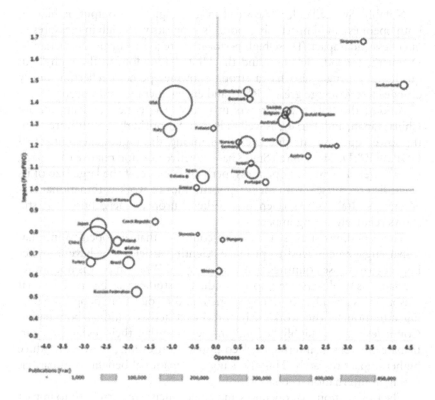

Fig. 7.1   Scatterplot of national citation impact compared to measures of national openness to international engagement

More researchers do not necessarily increase attention to published work. It is possible for more researchers to turn out more papers, but this does not necessarily mean that the work is of better quality or of equal importance to the scientific community. Some countries do much better than others at producing high-quality science.

The scatterplot shows three data points: (1) the X axis shows the openness index of a country based upon internationalization and mobility, (2) the Y axis shows the impact of a country's work by graphing the fractional FWCI of a country's publications, and (3) bubbles whose size is proportional to output (number of publications using fractional counting). The top right quadrant shows those countries that are both open and have a high fractional FWCI.

Notably, Switzerland, while small in geography and output, is high in both openness and impact. Singapore also appears very high in measures of openness and impact. These high performers are joined by the Netherlands, Denmark, Ireland, Belgium, and the United Kingdom in the high/high quadrant. Portugal also has a strong showing, perhaps reflecting policy changes to encourage greater R&D and engagement within Europe.

Among the lowest performers in terms of openness and impact are China, Japan, and Turkey, as well as Russia. Surprisingly, South Korea is in the lower quadrant despite spending among the highest percentage of GDP on R&D. The United States has a positive position relative to impact, but a lower showing on openness, perhaps because of the large size of its scientific enterprise—with many opportunities to work with national collaborators. Italy is less open than other European neighbors, but still shows relatively strong impact.

These findings suggest that those countries that are open to international engagement tend to produce scientific articles that have a higher impact than those countries that are less open. While impact is not always the same as quality (since poorly conducted studies can be cited for that reason), it is an indicator of engagement and recognition: people are paying attention to the work being produced across national boundaries. Countries that are highly "open," in the sense that their researchers participate actively in international co-publication events, tend to produce higher impact research. This also suggests a national benefit from participating in international collaboration.

The observation that openness and engagement are correlated to impact does more than simply confirm findings that show citation gains for international collaboration. It suggests that scientific mobility and connectivity may be factors in higher impact, which in turn encourages higher quality work. The European Union has built its R&D funding programs on the premise that collaboration may raise impact, and it appears to have borne fruit.

The finding of a relationship between openness and impact also suggests reasons for recent anomalies in the shifting positions of countries in terms of scientific output and leadership. Those nations that are less "open" appear to be lagging in terms of impact. Japan, in particular, has seen output and citation impacts remain flat since 2000. Japan is also among the least internationalized of leading nations. The lack of international engagement may be dragging on Japan's performance. Writing in 2010, Adams (2010) noted that Japan has a well-established research enterprise and world-class universities: "…[so] it is puzzling to the observer that the average rate of citation to its research articles in the internationally influential journals ... is significantly below...[other nations]." Lack of "brain circulation" may be the answer to this puzzle. In contrast, small- and medium-sized nations with enhanced global engagement have seen significant jumps in impact (Leydesdorff et al. 2014). Notable among these—and in addition to the well-known leaders of Switzerland, the Netherlands, Denmark, the United Kingdom, and Sweden—Singapore, Portugal, Belgium, and Austria stand out as countries that have increased their global reach and impact, with enhanced attention to their research. It appears that cross-boundary engagement and mobility have had positive effects in Europe, in particular. The location of large-scale laboratories and equipment may have an effect.

The United States holds several anomalous positions at the global level in terms of openness and impact. First, it has been noted that the United States has seen drops in percentage shares among highly cited publications. This is partly due to the increase of papers coming from many other countries, but specifically from China. In fact, despite the EU overtaking the United States in top 10% most highly cited publications, the United States was still leading the world in producing the top 1% most influential papers in science as measured by citations. (China does not produce a high percentage of highly cited papers.) This is the case even though, in percentage terms, the United States is less "open" than other leading nations.

The United States continues to attract scientists from around the world, as seen by examining the country of origin of many Nobel Prize winners from the United States—who, in 2017 and 2018, were all immigrants to the United States. However, the United States appears not to be sending a proportional number to study and work abroad. This may be because, compared to the other countries analyzed, the United States has a huge scientific workforce that collaborates and moves freely and frequently between its constituents states. The size of the US system in combination with a home bias to citations might also result in inflated impact figures in comparison to smaller systems.

The correlation between openness and citation impact is strong, also when controlled for by R&D funding and numbers of articles published. Countries with low openness and low impact include Russia, Turkey, Poland, China, Japan, Latvia, Lithuania, the Czech Republic, and, against expectations, South Korea (which spends a higher percentage of its GDP on R&D than almost every OECD country, including the United States). These countries are shown in the lower-left quadrant. Mexico performs well below what might be expected from the observed correlation between openness and impact of other countries. While an OECD country, Taylor (2016) argues that the lack of stable and sustained investment in Mexico has reduced the effectiveness of national spending.

Policy actions to nationalize research practices and reduce international engagement would appear to be antithetical to impact, and perhaps to creativity. While we cannot draw a causal relationship between openness and impact based upon this analysis, the initial indication is that "brain circulation" may be critical to bringing fresh ideas, enhancing creativity, and raising quality. Taylor (2016) argued that the scientific and technological prowess of nations is strongly related to their integration in international commercial networks.

REFERENCES

Adams, J. (2010). *Global Science in Perspective*. Conference Paper.

Archambault, E. (2010). 30 Years in Science: Secular Movements in Knowledge Creation. Science-Metrix. www.science-metrix.com/30years-Paper.pdf. Accessed July 2018.

Fukuyama, F. (1999). *The Great Disruption: Human Nature and the Reconstitution of Social Order*. New York: Free Press.

Harrison, C. E., & Johnson, A. (2009). *National Identity: The Role of Science & Technology*. Washington, DC: Osiris.

IMD. (2009). *IMD World Competitiveness Yearbook*. Lausanne: World Competitiveness Center.

Kumar, A., & Asheulove, P. (2011). *Report on World Publications*. Conference Paper.

Leydesdorff, L., Wagner, C. S., & Bornmann, L. (2014). The European Union, China, and the United States in the Top-1% and Top-10% Layers of Most-frequently Cited Publications: Competition and Collaborations. *Journal of Informetrics, 8*(3), 606–617.

Moed, H. F., & Halevi, G. (2015). Multidimensional Assessment of Scholarly Research Impact. *Journal of the Association for Information Science and Technology, 66*(10), 1988–2002.

OECD. (2014). *OECD Science, Technology and Industry Outlook 2014*. Paris: OECD Publishing. https://doi.org/10.1787/sti_outlook-2014-en.

Radosevic, S., & Yoruk, E. (2014). Are There Global Shifts in the World Science Base? Analysing the Catching Up and Falling Behind of World Regions. *Scientometrics*. https://doi.org/10.1007/s11192-014-1344-1

Science-Metrix. (2009). *30 Years in Science: Secular Movements in Knowledge Creation*. http://www.science-metrix.com/30years-Paper.pdf. Accessed Sept 2018.

SCImago. (n.d.). *SJR — SCImago Journal & Country Rank* [Portal]. Accessed 2014, from http://www.scimagojr.com.

Taylor, M. Z. (2016). *The Politics of Innovation: Why Some Countries Are Better than Others at Science and Technology*. New York: Oxford University Press.

CHAPTER 8

Local Innovation and the Global Network

The locations of knowledge centers—those geographic and institutional places where innovation is most likely to occur—are not evenly spread across the globe. There is no reason to think they should be, but as UNESCO (2016) reports, there is striking evidence of the persistence—expansion even—in the uneven distribution of research and innovation at the global level. When we look at the geography of clusters of capacity and knowledge, it is easy to see that there are heavy concentrations and that many of these concentrations are in regions within wealthy countries. Investment in R&D appears to remain concentrated in a relatively small number of locations within a limited number of countries (even though that number is growing, as we have seen). UNESCO (2016) points out that, in Brazil, 40% of government expenditure in R&D is spent in the São Paulo region. They further note that the proportion of concentrated government spending is as high as 51% in South Africa's Gauteng Province. Certainly, we are all familiar with the famous cases of Silicon Valley, California, and Bangalore, India.

Even where we see knowledge centers on a global map, we know that not all of these places are equally successful when it comes to attaining the various general goals of economic growth, environmental sustainability, public health, consumer protection, and so forth. (See Richard Florida (2005): The World is Spiky.) This chapter begins by acknowledging this disparity and diversity in terms of socio-economic organization and concentration as well

© The Author(s) 2018 141
C. S. Wagner, *The Collaborative Era in Science*, Palgrave Advances
in the Economics of Innovation and Technology,
https://doi.org/10.1007/978-3-319-94986-4_8

as the diverse goals of innovation policy for regional development strategies. Addressing the disparity is one of the central challenges of science policy in the twenty-first century.

The tenacity of geographic concentration of wealth is obviously not a new observation—it has been the driving question for much of economic inquiry. Why are some nations wealthy and others not? Writing in his book, *Industry and Trade*, in 1920, economist Alfred Marshall (2006) noted that economic activity was drawn to regions with a rich "atmosphere" of ideas. Amartya Sen (1999) writes about knowledge concentrations and their connection to democracy in *Development as Freedom*. Jane Jacobs (1969) argues that the growth of cities is based on a positive cycle of linkages among industries: the social and economic linkages among diverse activities generate and sustain growth. Diane Crane (1972) applies these concepts to scientific research in *Invisible Colleges*, and points to numbers of scientists and engineers within regions as key to regional growth. Richard Florida (2002) suggested that a creative class of people is required for a region to grow.

But here we focus on scientifically rich regions, and why some are globally connected and some are not. Clearly, even within the United States, some regions have the benefits of a high standard of living, a great deal of economic vitality, and a scientific system that is fully engaged at the global level, while other regions do not have these features. This question of regional concentrations of capability has received a great deal of attention, and the literature on regional knowledge, spillovers, and economic growth is rich and voluminous. This book does not claim to address the span of this discussion, rather to discuss how to use networks to disseminate knowledge.

Despite the many studies of location, innovation, and knowledge-based industries, no general theory of the relationship between knowledge and location exists. The degree of localization of knowledge appears to vary by region and by industry or sector. The semiconductor industry is a well-known example of localization: Silicon Valley has been the epicenter of key research and business activity in that sector, and through a series of feedback loops and virtuous cycles, it has drawn in people from all over the world to contribute their talents to world-class information technologies. At some unknown point, the collection of talent, funding, and other resources gets locked-in and positive feedback sustains the sector, as Dolfsma and Leydesdorff (2009) described.

Hundreds of regional government programs over the past 30 years have aimed to create "Silicon [fill in location here]" outside of California, but these have been largely unsuccessful. The conditions for regional agglomeration cannot be constructed and imposed. They emerge. The emergent features are not unlike the ones seen in the global network of science; they include initial conditions of agglomeration of people and resources, and feedback loops that enrich these and then attract others. Some of the factors contributing to regional economic agglomeration remain dimly understood, although they do seem to be tied to the locations of educational institutions, an educated populous of creative people, low barriers to business development, and transferrable capital. But an additional dynamic of culture and communication is surely at work—even though these are difficult to measure. The culture of cooperation can be nurtured in ways that enhance the links to the global network of science.

The rapid growth of the global network of science has implications for economic growth policies. On one hand, the global system offers enhanced opportunities to access knowledge and, possibly, new markets, and the system is "open" to new entrants who have the resources and are able to make connections. It offers opportunities for firms, researchers, universities, entrepreneurs, and civil society to access knowledge creation and use. On the other hand, the rapid pace of change means that developing countries cannot borrow a roadmap of development created by those who tread twentieth-century pathways toward growth. "Rapid developers" like South Korea, whose story of government-led development is inspiring (see Amsden 1992), cannot offer a workable model in the twenty-first century, since the conditions under which these nations developed are no longer operating. Today, developing regions must craft flexible policies and plans responsive to their own needs and capacities. This challenge requires strengthening each of the functions of the knowledge-based system to make it attractive partner in the network.

Economists differ as to whether institutions or technology are more influential on per capita income growth. Chris Freeman (1991), citing the World Bank, concludes that intangible investment in knowledge accumulation is more closely tied to growth than physical capital investment, which was, at one time, believed to be the essential element (Freeman 2002; Dahlman and Aubert 2001). Moreover, most economists agree that technological mastery grows from a knowledge base that first encourages the development of a cluster of capabilities that grow stepwise, and second, fosters a more complex, evolutionary process of "technology deepening." This growth process involves increasing the complexity of the production processes and adding value to and differentiating products based upon local market feedback loops. This process is inherently local, and often involves experiential learning.

The importance of a concentration of knowledge (in people and in institutions) as a key anchor for regional economic growth—especially for technology-based sectors—is widely acknowledged. Universities have recently further emphasized their "third mission" of knowledge dissemination to aid economic growth. Cities have begun creating their own innovation plans, with the idea of enriching a local innovation ecosystem to create high-quality jobs. Neil Coe et al. (2004) argue that the "the interactive effects" of knowledge exchange done face-to-face contribute more to regional development than factors known as "inherent regional advantages" (p. 469). This theme of interaction and multiple channels of communication as key features of regional development are emphasized in the evolutionary economic geography literature of Koen Frenken et al. (2009) and others.

Economic geographers argue that successful regions display a greater capacity for collective action and the ability to learn (MacKinnon et al. 2009). Studying the semiconductor industry in the 1990s, Almeida and Kogut (1999) observed that knowledge held by skilled individuals—often called tacit knowledge—within a region may be a key part of regional capability to support and sustain innovative industries. They are not alone in noting the importance of a skilled workforce, as they themselves report. Studies on innovation point clearly to the importance of educated people, since much of the relevant knowledge needed for innovation is not easily written down. The ability to learn new skills has been called "absorptive capacity," and indeed, this concept appears to be highly correlated with regions that can adapt to shifting knowledge sources and factors of production.

The importance of knowledge concentration at the regional level, and the communication and interaction with the global network, suggests that much of the policy made for the national innovation system may be misdirected. The "national model" was developed by Lundvall (2009), Freeman and Hagedoorn (1992), and others by observing highly developed countries operating in the twentieth century. Just as we see with the global network of science, with innovation, it is not clear that national boundaries or national-level policies are the determining factors in the interactive communications that transfer knowledge. Indeed, a more flexible framework may be needed, one that encompasses the five essential features and functions required to implement a knowledge-based growth policy within regions, presented in this chapter.

The framework proposed, developed in cooperation with Sara Farley, explores those conditions that best enable the development of a regional knowledge-based system that creates and absorbs, and then exploits and retains, useful knowledge. It is nicknamed "THICK" to represent the specific functions highlighted within it: technology, human resources, institutions, communications, and knowledge, as well as to indicate the kind of development needed on the ground: a thick, dense network of resources and communications. Toward this end, the model leans more heavily on the communication functions needed for interaction and those supporting functions of growth, rather than the institutions that may sustain it—as important as these may be to ultimate success. While the economics of innovation is a large literature, this chapter has a more modest goal of presenting a framework for applying what we know about network dynamics that will enable a region to take advantage of the global network of science.

Any region seeking to access and take advantage of the global network must begin with a regional inventory of knowledge capacity and the key challenges facing the region.

CAPACITY INVENTORY

The capacity of a region to enhance connections to the global network of science combines regional challenges with available knowledge—embedded in people, institutions, and technology, and the communications functions that connect these factors at multiple levels. These factors contribute to what is called the "learning economy" in a "learning region," as Kevin Morgan (2007) calls it.

Morgan suggests that the first proposition for the learning region is that innovation is an interactive process between firms and the basic science infrastructure. As we have seen, the basic science infrastructure is now a global phenomenon, and thus, firms or other problem-solving entities need to be able to source knowledge in "the cloud." But as we have also noted, global collaboration is a reciprocal process—in order to join in, one must have something to offer. The offering can be as modest as an interesting problem. But often, regions will have embedded technologies (and perhaps the underlying scientific knowledge) that can serve as the basis for collaborative linkages. These technologies are the first features to consider in an inventory of regional capacity. The concept of technology should include the tools, as well as the knowledge to use them (tools without knowledge are artifacts). As the previous section explained, access to the global network is neither the exclusive nor the sufficient component of knowledge-based growth. The technology inventory examines the available technological resources that can provide the basis for research collaboration. If a region is strong in agriculture, it may have the tools to initiate a bioproducts research program. If a region already has chemical processing plants, a case can be made for biochemical research investments. The regional capacities can be mapped using metrics such as patents, sales, corporate registration, R&D reports, strategic alliances, investments, tax reports, and labor rolls. This mapping provides the resources that can be contributed to a collaborative project.

Kevin Morgan suggests a second proposition about the learning region: that innovation is shaped by a variety of institutional routines and social conventions. To join the global network, we add that trained practitioners must be available within institutions to take part in collaboration. Trained scientists and engineers often know one another, and to have developed a level of trust. If not, this is a first step. Local knowledge is a strategic resource. Working with regional scientists or engineers to identify skills is a crucial step. There may be a case made for additional training by local practitioners if they feel they are unable to offer an attractive team to global collaborators. Further, it has been noted that often the most elite of these practitioners are engaged already in global, not local, problem-solving (Barnard et al. 2010). Locally engaged practitioners who do not have the ability to link to colleagues in their field can lose touch with the latest advances. (This sometimes happens where scientists or engineers are over-burdened with teaching loads that prevent them from participating in research.) This dichotomy of behavior is not uncommon in regions, and it can contribute to weak absorptive capacity. Incentives and tools must be in place to encourage both local and global links and provide local feedback loops and processes that render science useful to address local challenges. Policymakers should consider fostering linkages between domestic research and training institutions and their local counterparts.

This proposition of surveying local scientists and engineers to find out about their research, their interests, and their goals also has the benefit of building an inventory of "tacit" knowledge. The published record is highly revealing, but it cannot fully represent the knowledge that Ikujiro Nonaka and others (2006) point to as "that which cannot be taught or passed on." Many times, this knowledge includes an intangible understanding of local conditions that influence how well a scientific or technological solution could transfer to regional use.

Examining the functions needed for basic growth, and then determining the scale of a "base station" needed to link local needs to global capabilities, can create a pathway to networking. At times, the local "base station" can be quite rudimentary if global links fill in functions that can then be deployed locally. For example, product and process standards and metrology may be the most basic and therefore the most important function to establish S&T-based growth. Connections with inter-governmental groups such as UNIDO or links with standards organizations in other countries and through regional initiatives, such as has been done with the Quality Infrastructure in the East Africa Community (EAC) Initiative, can augment a standards and metrology institute in the local context, provided links are robust enough to benefit from capacity elsewhere. This EAC regional effort launched the first-ever East African Community Quality Web Portal, a resource shared by the five EAC countries. Many globally operating aid agencies were in place to help with similar efforts, although local governance and direction is critical to the success of these kinds of initiatives.

Collaboration—connections and communications among the different parts of the system that diffuse knowledge and enable learning—is the core part of any effort to link to the global network. Collaboration can take the form of international scientific or technological projects, research extension services (such as those between agricultural researchers and farmers), conferences and demonstration opportunities, online networking opportunities and resources (including web-enabled collaboration platforms like ResearchGate), and other opportunities—both formal and informal—through which people share ideas and strive toward joint problem-solving. Communication can include strategic efforts to tap into local or distant capabilities in product standardization, trade regulations, capital investment, technical advice, and supply chain implementation. Information and communications technologies have become an essential part of any knowledge-based system of innovation through which individuals and institutions connect and collaborate. Many regions have been focusing on improving their ICT infrastructure and usage rates. In some cases, developing countries have been able to implement leading-edge technologies such as mobile telephony with good success. For example, Uganda's AppLab, a joint venture of the Grameen Foundation and the cellular provider MTN, leverages the existing base of Village Phone Operators (VPOs) to bring information services to the remote villages via mobile devices, providing opportunities for micro-entrepreneurs to enhance their existing shared phone businesses with new value-added services. These kinds of resources can be provided by policy institutions, and they can often create the critical component to enable collaboration.

Knowledge—the embedded, underlying know-how either written as technical data and reports, or as guidelines and procedural documents, in regulatory and legislative code, in intellectual property, or embedded in indigenous knowledge—adds value and enables progress. Embedded knowledge is more obvious in advanced countries that have patents and journals that record knowledge for transfer. Regions also have embedded knowledge within local universities and industries and sometimes in local practitioners, where knowledge of local conditions and markets can be important for technology adoption and use. Local knowledge about botany, soils, seeds, and weather and water patterns can be critical to success in agriculture and energy production as in other key sectors, like health. The inclusion and validation of local knowledge in local industry's technological progress is perhaps the greatest contribution of a more holistic conception of knowledge and its application.

This framework of assessing technology and local knowledge tracks closely with the method used in the Inter-Academy Council Report, "Inventing a Better Future" (2004), in which the basic functions and capacities needed to use science successfully for industry-led economic development are highlighted, along the same lines of technology and embedded knowledge, institutions and infrastructure, social capital, and learning. With its emphasis on functions and features, the framework presented here is conducive to aiding the development of decision-making and investment within industry as well as government. It offers a human-centered approach to distilling S&T capacities at the firm level and the institutional level. It enables short-term action as well as a long-term planning. It creates the conditions necessary for local sustainability. It emphasizes the need for local networks. It focuses on building cross-cutting incentives. It draws upon resources and functions wherever they exist, rather than recreating them locally.

The approach suggested here is also supported by ideas about innovation found in the development literature. Going back again to Kevin Morgan's learning regions, he recommends a bottom-up approach that is demand-driven—as is suggested here. In our interpretation, the "demand" is created by local problems and challenges—what Morgan calls priorities for action. The aspiring learning region is called upon to "strengthen dialog between firms, and create regionally-based capabilities for research for technology diffusion" (p. 497), a similar process that we recommend.

Refining the THICK framework took place through an iterative process that extended over four years. Experts within Africa provided input to the construction of the THICK methodology at a World Bank-convened November 2007 workshop in Maputo, Mozambique, that drew together senior level S&T policymakers and other stakeholders from 20 African countries for consultation. The assembled experts noted: "Within the

African context it is important not to lose perspective and to always fully engage with rural communities, particularly with regard to initiatives centered on indigenous knowledge. The approach must always be innovation by the community for the community" (World Bank Institute & Finnish Embassy 2008). Subsequent review of THICK occurred at a 2008 meeting among industrial leaders and again during Uganda's National Science Week meetings in Kampala in 2009 and 2010 that assembled a broad and diverse constituency of policymakers, researchers, entrepreneurs, professors, and civil society. Additionally, a World Bank peer-review committee assessed the methodology and the resulting Uganda THICK review in 2010.

   With the vetted THICK framework to guide our inquiries, Sara Farley and I conducted a series of case studies and interviews in Uganda to test this model for application to S&T policy planning. The next section summarizes the application of the THICK model to support Uganda's planned efforts to build a knowledge-based economy.[1]

APPLYING THE THICK METHODOLOGY IN UGANDA

The THICK study of Uganda took place in three parts: (1) a review of existing literature and data on S&T investments in Uganda for each of six industrial sectors,[2] (2) a series of interviews with businesspeople, scientists, engineers, and academics knowledgeable about the subsectors pertinent to the industries, and (3) a workshop with government and academic

---

[1] The World Bank Institute (WBI) defined a knowledge-based economy as consisting of the following: (1) An economic and institutional regime that provides incentives for the efficient use of existing knowledge and the creation of new knowledge and entrepreneurship, (2) an educated and skilled populace that can create and use knowledge, (3) a dynamic information infrastructure that can facilitate the effective communications, dissemination, and processing of information, (4) an effective innovation system comprising a network of firms, research centers, universities, consultants, and other organizations that can tap into the growing stock of global knowledge, assimilate and adapt it to local needs, and create new knowledge and technology (Dahlman and Aubert 2001).

[2] The choice of case study subjects was made in consultation with the World Bank and with country representatives to ensure that the selections constituted examples of sectors that could benefit from an improved S&T infrastructure and capacity. The subsectors were chosen based upon relevance to national development needs, availability of data, and the presence of companies that could be approached for interviews. Interview subjects were chosen based upon their depth of knowledge of the scientific or technical aspects of the case study subject. Efforts were made to interview people within industry or professional associations. Priority was given to firms that were owned and operated by local businesspeople. Foreign direct investors were also interviewed when their input would increase our understanding of the sectors' dynamics.

experts as well as other expert advisors to validate findings and explore potential recommendations. The sectors and subsectors were chosen based upon their importance to the economy. For each of the cases, an effort was made to understand the S&T needs of that sector; inventory the knowledge available to that sector from local sources; examine the lines of communication between scientific or technical centers, academia, civil society, and industry; and identify the limitations placed upon industry as a result of those functions and capabilities that are scarce or absent.

The THICK methodology facilitates analysis across sectors (government, private, academic) and assesses Uganda's economic development in a completely different way. THICK is a complementary method to a value chain approach. Value chains encapsulate the sequence of steps, flows, investments, actors, and relationships that characterize and drive the process from production to delivery of a product to the market. In-depth value chain analysis considers several questions that can be ascribed to the five THICK dimensions: (1) What are the pathways from source to each end-market? (2) Who are the most important actors within the value chain and how do they behave? (3) How does a particular value chain connect to others, and what possible synergies exist? (4) In what ways is the value chain regulated from outside or self-regulated? (5) What is the institutional framework of the value chain (e.g., producer or trade associations)? The THICK methodology differs from the value chain in that THICK assumes non-linearity in the system—it crosses lines between the public, academic, and private sectors to identify the core capabilities that the industry needs to use science and technology as a catalyst for applying technical knowledge.

Technology Resources

The THICK analysis of Uganda's six key industrial sectors revealed that some strong technology applications can be found in Uganda's industries. The technical content of industrial production has increased over the past decade, largely through foreign direct investment encouraged by government. Case studies and analysis revealed three categories of technology in use. These technologies are aiding the efficiency of business and making it possible for some companies to conduct research and development. These technologies are:

1. Processing technology: use of numerically controlled machines and computer-assisted design software to increase the efficiency of production in manufacturing.

2. Computing: use of computers for supply chain management to track production, as well as to conduct accounting that is in line with international standards, to track products, and to store data.
3. Mobile telephony: use of mobile phones to keep staff in touch with each other, to trace where products are over the course of the supply chain, and to connect researchers to one another.

A number of technologies could be used profitably if they were upgraded and implemented in the industries we studied. Among these are:

- Numerically controlled machine tools for cutting materials that would aid many kinds of manufacturing projects; they would be particularly helpful in the agro-processing and transport sectors;
- Packaging and plastics of all kinds to create the kinds of packages that add value to agricultural products (such as breathable plastics for vegetables or foil wrap for coffee beans) and for products deriving from the ethnobotany sector;
- Logistical support (software and electronics hardware) through computer-based inventory tracking.

In several interviews, firms reported upgrading to ICT-enabled transport, logistics, and supply chain management systems. In these cases, the firms reported a host of benefits, including enhanced competitiveness, efficiency, and the pairing of related processes that were not previously linked.

Within the transport and logistics sectors, technological challenges are many and include securing fuel-efficient vehicles, dealing with poor road conditions, and tracking the production and distribution of products. The benefits gained from accessing and using technology to enhance firms' productive capacity are not spread evenly across firms and sectors. The gaps between firms in terms of access to software, hardware, and network connections create bottlenecks for the dissemination of knowledge about technologies that would aid companies. This is particularly true for small companies and independent farmers. The lack of basic infrastructure in reliable electricity, wired telephony, and broadband Internet hinders scientific development and technological adaptation. These gaps impinge upon firms' ability to link and connect and, thus, access useful knowledge.

Smaller firms are disadvantaged because they cannot afford to put technology into their business processes like larger firms. The review of the case studies revealed other significant gaps in access, use, adoption, and adaptation of technology resources. One of the most prominent gaps is that very few firms use technology throughout the production process. Most cases of technology usage in industry occur at only one point in the process, not as a continuous interaction throughout production. This problem results in efficiency losses throughout the process. In particular, there were very few cases where software linked production with raw material inventory, or with the distribution process—a part of supply chain management common in most manufacturing centers in other countries.

Human Resources

Ugandan industry representatives agreed on the need for increased opportunities for industrial internships and training. Beyond the obvious advantage of such arrangements in terms of refining the skills of employees to respond to the rigors of a particular work environment, industrial attachment appears to be increasingly common in several sectors as a means to compensate for inadequate training in the formal education sector. Specifically, in nearly every industry profiled, there was recognition of the need for improved vocational training and a willingness to work with local universities to improve training. Of the sectors we examined, energy and ICTs were more likely to have formal training opportunities available than transport, agro-processing, or ethnobotany.

Several companies reported existing plans to offer significant improvements in their training programs. University enrollment represents just 4% of the age cohort in Uganda (Uganda National Council for Higher Education 2004). Tertiary education remains out of reach for all but a tiny sliver of the population. Certainly, access to higher education needs to be improved overall. Within the universities there are well-trained faculty and good-quality students that can help build additional resources. At primary and secondary levels, national education policies and programs have focused on keeping students in school in order to raise overall literacy rates. This policy of access should be balanced with efforts that ensure the brightest students have access to tertiary level education and paraprofessional training, particularly in critical technical areas.

Despite the country's achievements in human resources development, the particular challenges related to S&T human resources development are substantial. There are not enough training opportunities to meet industry's demand. Present rates of Ph.D. production in the sciences—25 Ph.D.'s annually—are not sufficient to replace retiring professors, let alone respond to the raft of S&T needs demanded by a fast-growing, agricultural economy. The mismatch between skills needs and human resource availability appears to be particularly true for the field of software development, where there is a short supply of local computer engineers who can write or customize software to respond to industry demand. Most software operating in manufacturing is purchased from large international suppliers. While this is not a drawback in itself (since purchased software builds capability), the software products must be used off-the-shelf, even when a customized product would be more advantageous.

Regarding key S&T skills of Ugandan graduates, a number of industry representatives noted that there is a lack of some basic technical skills, such as electronics repair and computing, as well as a lack of professional certification for specific capacities, such as in chemistry. Several industries reported the need to offer training to technology users, such as short courses for local farmers and manufacturing workers on new technologies and ICTs. Practical experience is systematically weak over the course of formal training in most sectors examined. Training is predominately offered by companies with a large international reach, leaving out the small- and medium-sized enterprises (SMEs). Many SMEs operate under the assumption that they do not have the resources to integrate industrial attachment opportunities for students into their business model. In all fields, greater attention is needed to develop institutional mechanisms that integrate the voice of the private sector and the needs of industry in curricula development and reform for all levels of education, from basic, to vocational, tertiary, and graduate levels.

Institutional Resources and Infrastructure

Uganda's ability to harness S&T for growth and development needs to be supported by institutions. A wide variety of institutions bear on a country's ability to use, adapt, diffuse, and generate knowledge. This may include those offering services related to importing raw materials, customs and tariffs, metrology, standards, testing, and quality assurance (MSTQ), human resources education and training, law, policy, regulation, finance, and infrastructure. The THICK methodology offers a narrower view of institutional resources than other models do, highlighting only the most influential institutions for the application of S&T. Here, institutional resources include the following: (1) key S&T-related ministries and relevant parastatals, (2) standards setting bodies, (3) metrology testing centers, (4) educational and training organizations, (5) research institutions, (6) donor institutions, (7) incubators and science parks, (8) producer associations, and (9) financial institutions.

The existence of Uganda's National Council on Science and Technology signifies the commitment to prioritize S&T issues within government. The council's mandate includes coordinating S&T policy across government, academia, and industry. This institutional resource provokes dialogue across ministry lines, stimulating national-level dialogue on S&T-related reforms, though such dialogue may be considered nascent. For their own part, sectoral line ministries undertake planning on a host of issues, increasingly emphasizing the important role of S&T with respect to their sectoral domains, suggesting an institutional appetite for greater coherence around S&T issues.

A second positive finding resulting from the THICK analysis of Uganda's institutions concerns education and training institutions. Positive achievements toward increased access to education at the primary and secondary levels and exploding enrollment at the tertiary level demonstrate that the demand for education is high and institutions have been growing as a result. Government initiatives to earmark federal scholarships for S&T bode well for the emergence of larger cohorts of trained S&T graduates.

The creation of a Ugandan ICT government agency offers a third positive finding. This institutional resource is configured to support the development of the infrastructure, skills, organizations, and partnerships needed to maximize the contribution of ICTs in industry, learning, and governance.

The existence of business research and development institutions constitutes a fourth welcome discovery. Uganda's Industrial Research Institute, for example, engages in activities designed to facilitate the "rapid industrialization of Uganda by identifying appropriate and affordable technologies that will enhance adding value to local products so they can be processed for national, regional, and international markets."

The THICK analysis revealed that, despite consensus that modernization of key sectors depends upon improving the ability of people, firms, and institutions to access and apply knowledge and technology, some institutional barriers exist. Specifically, restrictions related to importing, procuring, accessing, distributing, and integrating knowledge and technology from regional or global sources present obstacles. Among the disincentives frequently bemoaned are high tariffs on imported technology, despite the fact that, according to African enterprises, importation of machinery is the most common route toward technological upgrading. Protracted delays due to burdensome customs or procedures constitute another institutional constraint impeding Ugandan industry's access and use of knowledge produced abroad.

Second, although institutions charged with various aspects of MSTQ are in existence, the competence of these institutions is highly variable. Information gaps often separate researchers, entrepreneurs, and teachers from needed insights regarding consumer and labor market demands, manufacturing standards, phytosanitary testing required for export, and so forth. For example, the institutions that are active in the MSTQ space in Uganda are Uganda National Bureau of Standards (UNBS), Uganda Industrial Research Institute (UIRI), the Uganda Analytic Laboratory (UAL), Chemiphar, and SGS. However, the existence of several institutions belies the fact that firms developing products for domestic consumption or export do not receive the kinds of services and technical assistance they need to enter markets in compliance with established quality and safety standards. A lack of accreditation, added to the other institutional constraints undermining these institutions' capability, reduces their practical usefulness to producers seeking entry into markets, domestically and abroad.

Where funding and connections to global knowledge (including through regional networks) are adequate, Ugandan researchers perform world-class, highly relevant research. Yet, according to the 2005 World Bank-supported Science and Technology Sector Profile of Uganda, "Essentially all research funding comes from external (donor) sources, for work on problems of concern to and defined by donors." Too few research institutions work on local problems and challenges defined by local stakeholders. In fact, the majority of the fewer than 30 universities in Uganda do not conduct research of any kind, the main exceptions being Makerere University and the Mbarara University of Science and Technology (MUST), plus a few of the emerging high-quality private institutions.

A challenge cited frequently by businesses and entrepreneurs relates to financial institutions and the lack of available financing for capital investment. For those firms seeking resources to improve their technological capacity, expand production, or enhance their processes through the application of knowledge, the lack of access to small loans discourages investment. Frequent reports of interest rates as high as 20–30% epitomize the barriers to knowledge investment that firms and entrepreneurs confront. Additionally, where opportunities for short-term skills upgrading and affordable technology acquisition occur, micro-financing rarely can be accessed. Improving the capacity of commercial financial institutions and micro-credit organizations to serve the needs of Ugandan industry is key.

Finally, there is a lack of science parks and incubators that create the conditions conducive for research, technology development, and business support services. Though science parks are not a quick-fix or one-size-fits-all commodity, when used effectively, they offer a host of benefits in terms of research and technology development, network facilitation, partnership creation, and innovation. Yet, with a few minor exceptions, the Ugandan knowledge system is largely bereft of these kinds of support systems.

Collaboration and Communication

Uganda's ability to access, adapt, and absorb knowledge is enhanced by opportunities to collaborate within and outside the country. Links between people, across sectors, and among public, academic, and private partners allow for the kind of knowledge exchange required for innovation. Among the strengths observed in Uganda with respect to its collaborative capacity, a perceptible shift toward greater openness and increased sharing of information appears to be taking place. With respect to S&T collaboration, the country's tremendous increase in mobile telephony constitutes a second noteworthy asset at present. The total number of telephone subscribers in Uganda increased from roughly 276,000 to 3.6 million between 2001 and 2007. As mobile telephone penetration improves, so too do opportunities for innovative uses of this technology as a means to link producers to consumers, researchers to users, and private entrepreneurs to public partners. Other key positive findings are:

- Increased computer usage
- Enhanced efforts to communicate the important role of science, technology, and innovation for society such as through the new National Science Week
- Occurrence of innovative efforts to reach out and connect industry with local needs
- Integration of computer systems (such as computer-aided design) into many industries
- The presence of extension services that connect researchers to producers
- Focus on regional S&T planning initiatives

Despite its commendable accomplishments, Uganda's progress toward realizing a robust communication capacity is limited by a number of continued and pressing challenges. Some of these challenges currently elicit attention but remain significant due to their complexity or due to the paucity of resources allocated to their amelioration, while others appear to persist largely unaddressed. We highlight all of these challenges.

Detrimental to any country's ability to tap local, regional, or global knowledge resources is the minimal existence of linkages between key actors within a knowledge or innovation system. In Uganda, few linkages exist between the research community, public research organizations, universities, industry, and users/civil society. As compared to the dynamic knowledge economies of OECD countries in which collaboration between industry and university is commonplace and in which public research organizations systematically link with both universities and industry, the institutional actors in Uganda remain in isolated silos. Across the case studies, interviewed representatives from discrete sectors (industry, university, or the public sector) described the frail or non-existent links with other sectors. The consequence of the lack of linkages is not just missed opportunities in terms of collaborative research efforts, however. Rather, in areas such as standards and testing, knowledge gaps in terms of market demands and industrial regulations restrict otherwise exportable products to domestic use.

A second challenge relates to channels for knowledge diffusion within the innovation system. Where relevant research takes place or technology development occurs, a lack of channels through which knowledge may diffuse reduces its occurrence. Interviews yielded frequent accounts of "industrially-relevant research" conducted at universities, technologies developed for which a perceived "need within civil society exists," and repositories of business development information sitting underutilized even when firms express strong demand for such information. The frequency of these occurrences suggests that the challenge of vastly enhancing explicit and efficient channels to distribute existing knowledge through the system is even more pressing than the challenge of generating new S&T-related knowledge. Need for knowledge is not sufficient to guarantee that the required information (in terms of technical know-how, market intelligence, knowledge embodied in technology, etc.) will make its way to those who need it.

A further challenge concerns the weak connections to consumers and international markets. These weak connections thwart Uganda in its efforts toward export orientation in the agroindustrial and ethnobotanical sectors. Weak connections to consumers and markets impose severe constraints on market development. For example, within the ethnobotany sector, market information critical for success in developing and exporting ethnobotanical products (such as cosmetics) includes information relating to sanitation and safety regulations, phytosanitary certification of the raw materials used, general quality requirements, health and safety issues, and information related to acceptable packaging, marketing, and labeling. In the absence of this information, small producers using rudimentary technologies to manufacture powders and crude essential oils restrict their sales to local markets and/or sell unprocessed raw materials for value-added processing to buyers empowered with a more savvy understanding of consumer and market needs.

Limited web access constitutes another challenge, weakening the overall communication capacity. Web access is essential as a tool to locate and use critical information in the course of research, product development, or marketing. Without it, researchers cannot upload their findings, publications, and materials onto the web to make them accessible to the global knowledge community. Further, web-based resources provide a platform to establish national presence and competence in emerging scientific areas or within market niches. Ugandans are less likely to elicit knowledge partners and collaborators over the course of their research due to this lack of participation in the online knowledge community.

Knowledge Base

The knowledge base in Uganda consists of six functions: (1) technical data and reports, (2) research funds, (3) technical development funds, (4) regulations, (5) laws governing safety and environmental protection, and (6) indigenous knowledge. While clean-cut distinctions between human resources and the knowledge base from which they draw and to which they contribute are difficult to construct, the measures included in the THICK methodology are designed to include not just the institutionally derived aspects of knowledge (e.g., laws, regulations research protocols) but the knowledge absorption dimensions as well (e.g., knowledge and enforcement of laws, technical reports, journal articles, technical advisory services). In this way, the THICK framework assesses both the depth of the knowledge base and the degree to which the existent knowledge base is tapped effectively. With respect to Uganda's knowledge base, six positive findings are highlighted, followed by seven of the key challenges confronting the knowledge base:

- Organizations within Uganda have created knowledge repositories for industry, such as the Uganda Industrial Research Center's industry web portal.

- Comprehensive sector analyses have been conducted, such as one on the oilseed sector (produced by the UOSPPA) and an ethnobotany sector assessment of natural ingredients for cosmetics and pharmaceuticals (produced by Natural Chemotherapeutics Research Laboratory).
- Regional knowledge organizations, through which governments, institutions, people, and firms collect and exchange knowledge, have formed.
- Platforms for knowledge partnerships between private sector firms and small and medium enterprises and/or entrepreneurs are emerging.
- Private sector research is occurring, such as that being performed by Bidco pertinent to palm oil.

The positive findings noted earlier suggest that Uganda benefits from an ability to contribute to, and elicit information from, the science and technology knowledge base, be it the local knowledge base, regional knowledge base, or global knowledge base. The positive findings should not overshadow the fact that knowledge exchange is not the norm within Ugandan industries, a challenge that must be addressed. Too few opportunities to share ideas, collaborate over technology development, and engage in cross-sectoral problem-solving are fostered or exploited. Thus, knowledge often does not circulate within distinct communities (university researchers, government ministries, public research institutions, etc.) or between them. With a greater emphasis on fostering such opportunities for knowledge exchange, the rate of knowledge acquisition and application could increase. Moreover, little attention is paid to the question of when and how knowledge (in the form of research, technology, advice, etc.) should be mined elsewhere as opposed to growing the capability for knowledge creation locally. Understanding where knowledge is located and when to prioritize scanning for it, accessing it, adapting and using it, constitutes an essential and yet weak aspect of the country's knowledge system.

A second challenge confronting the knowledge base stems from weak and uncoordinated legal frameworks for the commercialization and protection of innovations in the form of technology, new products, and new processes. Insufficient intellectual property rights (IPR) capacity in the form of little or no regulatory capacity, a lack of information regarding the existence or relevance of international rules and regulations, and a dearth of trained lawyers equipped to facilitate IPR-related agreements impinge entrepreneurs' proclivity to commercialize novel products and processes.

A third challenge requiring attention pertains to regulations and knowledge dissemination regarding MSTQ. The array of institutions charged with determining market demands for safety, packaging, content, and so forth (for both domestic and international markets), establishing standards, testing products at various stages of development, and enforcing adherence to quality and safety standards in domestically produced and imported goods is insufficient at present. While some of the weaknesses within the institutional array relate to other aspects of the THICK methodology, others bear on weak capacity to elicit critical information. As international standards for quality and safety are a constantly shifting target, Uganda requires near-constant investment in knowledge scanning, precision equipment, and the human capacity to run the required tests and conduct analysis. Weak knowledge flows between the standards and safety institutions and industry at home and abroad result in knowledge gaps. Knowledge gaps in terms of standards vary in severity, with some sectors bereft of standards all together, such as in the processing of natural ingredients into extracts, essential oils, and saps. These knowledge gaps weaken entrepreneurs' and would-be exporters' ability to develop their products in line with requirements for gaining entry into foreign markets. Until these knowledge flows are improved, reports of goods turned away at European shipping ports or refused by international buyers will be many.

Another knowledge-related challenge to be addressed relates to research gaps. Although pockets of high-quality research may be found in Uganda, serious holes in research capacity render the knowledge base thin and inconsistent. For example, though touted as an area of relative strength in Uganda's research community, agricultural research is characterized by underinvestment in certain areas key to the modernization of the sector as called for under the Plan for Modernization of Agriculture in Uganda (PMA) and the Comprehensive Africa Agriculture Development Programme (CAADP). Deepening Uganda's knowledge base will entail enhancing the appreciation of how scientific and technological know-how can be used (and from where it may be accessed) to solve urgent needs, like the 6% agricultural productivity yields called for in CAADP.

Finally, challenges related to gathering and absorbing market information impede prospects for industrial growth and development even as other knowledge resources are marshaled effectively for use in industry. Learning how to learn is essential here. Yet, systematically, Ugandan enterprises engaged in agroindustrial processing, ethnobotany, and other sectors reported that they did not know how to gather market information or ascertain customer demand, particularly in international markets. The unmet need for more intermediary organizations, similar to Uganda's Export Promotion Board, that manage knowledge flows between industry and markets is still a challenge. The choice of a realistic S&T model to guide the development of a regional learning economy operating in the global network is critically tied to success. This chapter argues that on the ground in Uganda and within other countries, the THICK framework shows that focusing on local growth of functions surrounding technology, human resources, institutions and infrastructure, and features of communications and collaboration, and embedded knowledge is closer to the needs of a region than is national policy. Rather than a sector-by-sector approach, a systems approach of applying these functions to local needs, building the features in concert and interaction, is more likely to result in local and sustainable growth. Local solutions can be crafted in reference to the specific functions and supported by global links efficiently introduced as needed.

References

Almeida, P., & Kogut, B. (1999). Localization of Knowledge and the Mobility of Engineers in Regional Networks. *Management Science*, 45(7), 905–917. https://doi.org/10.1287/mnsc.45.7.905

Amsden, A. H. (1992). *Asia's Next Giant: South Korea and Late Industrialization*. New York: Oxford University Press on Demand.

Barnard, H., Cowan, R., & Muller, M. (2010). Global Excellence at the Expense of Local Relevance, or a Bridge Between Two Worlds? Research in Science and Technology in the Developing World. MERIT Working papers, No. 51. Maastricht: United Nations University.

Coe, N. M., Hess, M., Yeung, H. W., Dicken, P., & Henderson, J. (2004). "Globalizing" Regional Development: A Global Production Networks Perspective. *Development*, 29(4), 468–484. https://doi.org/10.1111/j.0020-2754.2004.00142.x

Crane, D. (1972). *Invisible Colleges: Diffusion of Knowledge in Scientific Communities*. Chicago: The University of Chicago Press.

Dahlman, C., & Aubert, J.-E. (2001). *China and the Knowledge Economy: Seizing the 21st Century*. Washington, DC: World Bank.

Dolfsma, W., & Leydesdorff, L. (2009). Lock-In and Break-Out from Technological Trajectories: Modeling and Policy Implications. *Technological Forecasting and Social Change*, 76(7), 932–941.

Florida, R. (2002). *The Rise of the Creative Class*. New York: Basic Books.

Florida, R. (2005, October). The World Is Spiky. *The Atlantic Monthly*.

Freeman, C. (1991). Technology, Progress and the Quality of Life. *Science and Public Policy, 18*(6), 407–418.

Freeman, C. (2002). Continental, National and Sub-national Innovation Systems – Complementarity and Economic Growth. *Research Policy, 31*, 191–211.

Freeman, C., & Hagedoorn, J. (1992). *Globalisation of Technology*. MERIT.

Frenken, K., Hoekman, J., Kok, S., Ponds, R., van Oort, F., & van Vliet, J. (2009). Death of Distance in Science? A Gravity Approach to Research Collaboration. In *Innovation Networks* (pp. 43–57). Berlin/Heidelberg: Springer.

Jacobs, J. (1969). *The Economy of Cities*. New York: Columbia University Press.

Lundvall, B. A. (2009). Innovation as an Interactive Process: User-Producer Interaction to the National System of Innovation. *African Journal of Science, Technology, Innovation and Development, 1*(2–3), 10–34.

MacKinnon, D., Cumbers, A., Pike, A., Birch, K., & McMaster, R. (2009). Evolution in Economic Geography: Institutions, Political Economy, and Adaptation. *Economic Geography, 85*(2), 129–150.

Marshall, A. (2006). *Industry and Trade*. New York: Cosimo, Inc.

Morgan, K. J. (2007). The Learning Region: Institutions, Innovation and Regional Renewal, *31*. https://doi.org/10.1080/00343409750132289

Nonaka, I., Von Krogh, G., & Voelpel, S. (2006). Organizational Knowledge Creation Theory: Evolutionary Paths and Future Advances. *Organization Studies, 27*(8), 1179–1208.

Sen, A. (1999). *Development as Freedom*. New York: First Anchor Books.

UNESCO Science Report Towards 2030. (2016). https://en.unesco.org/unesco_science_report?language=fr. Accessed June 2018.

CHAPTER 9

Governing Global Science

The human condition is not improved by single facts of science. In their collection into a body of knowledge, joined to its careful application, we can stretch the boundaries of justice still larger, as David Hume exhorted us in seeking to use science to solve problems. The process of funding and supporting science results in an unequal distribution of knowledge. This is the nature of the system, but it is up to policymakers to see to the equitable distribution and dissemination of the results and benefits. This process of diffusion and distribution is facilitated by public policy, but, as we have seen, it no longer occurs at the boundary of the nation-state. Science into practice advances welfare, but that process does not happen on its own. In scientifically advanced countries, people and institutions are in place to translate science into practice, often with specific goals in mind, such as efficient energy, purer chemicals, and clean water. As a result of these goals and an investment in a knowledge-creating system, these countries are often more able to absorb and use scientific knowledge, and they are to do it more quickly than other places that have no function for sharing knowledge (sometimes called the *translational process*). In developing countries, some institutions are in place to aid in the translation of science into practice, but these institutions are weaker than in advanced countries, and the ones in place can be vulnerable to decline during economic downturns.

The role of these scientific knowledge translators—including policymakers, analysts, engineers, business entrepreneurs, and sometimes scientists themselves—is critical to advancements that improve welfare. To be

© The Author(s) 2018
C. S. Wagner, *The Collaborative Era in Science*, Palgrave Advances
in the Economics of Innovation and Technology,
https://doi.org/10.1007/978-3-319-94986-4_9

effective, the translators, interpreters, and distributors of knowledge require a clear and robust understanding of the dynamic system of science, and one that acknowledges not just the global network of science but also the global need for scientific knowledge. This chapter summarizes what we now know about the global network of science and discusses implications of these findings for policy and governance.

The tendency of the science system to evolve toward an elite structure, demonstrated in this book by the persistence of preferential attachment in the global network, is not a new observation, nor a new problem. As Derek De Solla Price (1963) wrote, "The hierarchy of good scholarship is extremely acute," meaning that the spots at the top of a pyramid—or the center of a network—are few and hard won. This feature of science, where just a few preeminent researchers reach the pinnacle of achievement and reputation, appears to be the natural order of science. This tendency is heightened by the network dynamic, where reputation drives the system toward concentration and information technology accelerates the process of identifying 'stars' within the network. This, in and of itself, does not present a problem, since excellent science can and does still get done. The question becomes one of how best to work within this system to disseminate knowledge and to improve outcomes—especially for building the public good.

Within the network, as we have seen, are nested hierarchies of scientists who have contributed to building their own structures—ones different from the political borders and organizations—that underlie the network. The self- and other-selection into the global network has driven scientific knowledge creation toward the global and away (to some extent) from the national-level institutions that have housed and governed it in the past. This presents significant challenges for these national institutions and program managers, as well as for other institutions and people, that depend upon science (and pay for and account for the spending) to advance human welfare and technological competitiveness.

Guidelines for managing science are formed at the national level with national constituencies in mind. This means that legislatures and bureaucrats allocate funds to government missions and to the growth of basic knowledge. For science policy, this has worked well for many decades, in part because institutions serve to achieve national, regional, and local goals. By doing so, projects and programs provide accountability within the political system. Institutions often collect data about spending, trained people, publications, applications, and other indicators, to mediate between those who conduct scientific research, those who "consume" that research in the form of economic growth and social welfare, as well as those who oversee it on the part of the public. The links (as we have seen) are difficult to disentangle for the purposes of direct attribution—for example, who is responsible for what part of a multi-authored, international paper?

Funding for scientific research—while only a part of the activities we call "science"—is the part that receives public attention for several reasons. First among these is the budgeting and cost that is part of R&D. Research is expensive; it often requires highly tuned equipment, highly trained technicians, and laboratory space or travel to unique sites. Second, it is risky and uncertain; R&D is likely to turn up "negative" results, requiring more experimentation than expected at the outset. Third, it is difficult for any single entity to appropriate the benefits—the results can be (and should be) shared and entered into the public commons for broader use by others. This latter feature is the one most often cited as justification for public investment.

Public spending on scientific research is generally divided between the application of scientific and technical knowledge to meet national missions (such as energy or health)—sometimes called *science-informing-policy*—and research into uncertain areas of inquiry at the frontiers of knowledge, where advances are believed to be possible and viewed as needed by the scientific community—sometimes called *policy-for-science*. The economics of these activities have been the subject of many studies, although the actual benefits surely differ within varying settings and governance structures. (Private sector businesses benefit from government-funded research, but for different reasons than benefits derived by the public for, say, national defense.) No single rule applies for allocating public spending on scientific research (with the possible exception of the OECD-suggested goal of 3% of GDP contributed to all of research and development spending, a goal which serves as more of an aspiration than a benchmark, and which few nations attain).

As we have discussed, more countries of the world are committing national funds to scientific research, and many intergovernmental groups such as the World Bank and the United Nations have increased support to these activities. National funding of and need for science and technology research are not being reduced by the global network of science—far from it. National, regional, and local needs remain an important part of the demand for scientific research. The question is how to construct policy to tap into the network.

Traditional policy actions will likely continue because national agencies and ministries meet more goals through science policy than simply funding science. In addition to these missions that are part of science investments, governments need guidelines designed to govern and distribute the global goods within the network of science. Enhanced national policy is needed, together with some reorganization of international organizations and associations. The challenges are to fashion policies within existing institutions at the international and national levels, at administrative levels, and at the state and local levels, designed to better interact with and exploit the use of networked (as opposed to hierarchical or linear) systems. The goals are greater efficiency and effectiveness of public spending, and enhanced justice for those lagging behind who could benefit from better accessing the global system. As one scholar put it, the challenge is "[h]ow to turn network knowledge, as it develops, into practical, useful information."[1]

Creating policy guidelines requires considering the organization of governing groups and their goals, the goals of scientists, and the ultimate usefulness of science. The global network of science—much more than national systems—offers the possibility of expanding the horizon of justice and inclusion. Elinor Ostrom (1990) described a system wherein knowledge is a "global public good." The global network of science, on account of it being open, accessible, and dynamic, can be discussed as a resource within Ostom's framework. Ostrom's breakthrough analysis focused on "common pool resources," which can face problems of congestion or overuse because they are subtractable, such as resources like open-sea fishing fields. Unlike many of Ostrom's examples, knowledge is neither subtractable—one person's use reduces the portion available to the next user—nor is it easily excludable. Indeed, for much of scientific knowledge, sharing provides *network effects* (not "network" in the same sense we have been using in this book), where a good increases in value to the extent it is adopted by the next user. To the extent that we share common scientific or technical knowledge, common action becomes much more likely to occur and with better outcomes.

In Ostrom's framework, scientific and technical knowledge have features closer to a pure public good in that its value is not diminished by use—in fact, knowledge is generally enhanced by shared use. The openness envisioned for a broader policy toward participation in and support of the global network of science is not simply one of open access to the results of science—here referring to the movement of scientific publications from behind "pay walls" and into freely available sites on the Internet—but openness in the sense of active inclusion of many more

[1] O'Toole (1997, p. 48).

practitioners into the global network. For want of better vocabulary, openness here means inclusiveness—where the results of inquiry can be accessed and governed by the users themselves, and where producers and users can participate in disseminating the results to those who can productively use the knowledge.

OPEN SCIENCE

A vision of open science—in which knowledge is shared, disseminated, and translated into action—is one where science is not an elite and exclusive, obscure activity (an elite system was described by Vannevar Bush (1945) in "Science, the Endless Frontier"), but one in which useful knowledge is widely available and actively applied to improve human conditions (Mokyr 2002). This vision is one in which people participate in research, development, dissemination, and use in many more ways than has been practiced in the past. This can include practices such as sharing health data across mobile devices, offering real-time crop advice to farmers, examining water quality and providing that information to consumers, and monitoring seismic activity—all can be done using communication devices that remove or reduce physical distance. This process can also include more "citizen science," such as tracking wildlife migrations, astronomical phenomena, and environmental conditions, all reporting back to a centralized, online, and public clearinghouse.

An example might be a much updated and connected version of the United States Agricultural Extension Service, created in the nineteenth century to ensure that the best agricultural science and technology was disseminated quickly, effectively, and widely to users. Much of this service was provided by extension agents traveling many miles and working with farmers to provide know-how, and to observe problems in the field and report these back to practitioners in the Land Grant colleges that conducted basic and applied research on related problems. The extension of knowledge, and the feedback loops, created a long, unfolding revolution in agricultural production that completely transformed life in the twentieth century.

A similar revolution is available now, taking the form of a knowledge utility. Here, we can draw upon a similar vision as that embodied in agricultural extension, but a knowledge utility relates to agricultural extension as perhaps a three-dimensional chessboard might relate to a game of checkers. The ultimate goal is to connect many more people to the results of scientific research, with broader outreach for funding and knowledge dissemination, in ways that would facilitate greater use and application. This recommendation envisions a governance model dedicated, not just to the support for good science, but also for its dissemination and application for public welfare.

The challenges for realizing this vision exist at three levels of government: federal, state, and local, or perhaps better separated into macro, mezzo, and micro levels. The author is most familiar with the US governmental model, but in all political systems, some aspects of macro, mezzo, and micro governance structures exist. The challenges of governing the global system differ at each level in terms of ability to influence network structure and operation as well as to take advantage of the benefits of the results of networked science. The goal of broader participation and dissemination is discussed at each of these levels.

Given the public interest in and needs for the results of scientific research, governments retain a keen interest in funding it. However, the global network is not well aligned with national or local goals. This means that a combined approach of funding science for national missions and scanning the globe for critical knowledge that can be linked back to local and regional needs is the broad goal. This certainly does not mean that nations stop funding science—simply that they adjust the way the funding and knowledge dissemination is done, moving away from a competitive posture and toward one of integration.

Governing science as a global system can potentially provide richer resources than could have been created within a national system. That much is clear. However, few good models of success exist as guides for governing the network of global science. Governmental functions (e.g., tax policy and revenue collection) and services (e.g., the post or social security) are made available to the governed because government exists—these are the functions of government. These are public functions, and they are governed according to rules of good governance. A linear relationship exists between their function—from planning to execution—and the impact—from execution to evaluation. The linear relationship gives government strong control of their operation. The same model does not apply to science or to global networks.

Scientific knowledge does not fit earlier models of governance—where government provides the knowledge needed to support a knowledge utility. We are only in the earliest stages of viewing this role as within the purview of government. Moreover, science is an elite activity supported within democratic systems, which brings internal conflict associated with it.

However, governments often fund other elite activities, such as art, music, and dance; we understand that only a few people have the skills and talent to contribute, even though many can benefit. Science has similar features to these functions, except that these others are done for their own sake, whereas science is done to benefit national missions and social welfare.

The global network creates highly functional knowledge, which creates links that are not highly efficient at knowledge dissemination. We cannot ask scientists to become expert extension agents—this function should be provided by government. Science operates by its own internal reward system. That reward system, again, works well for science, but not necessarily for knowledge sharing. Knowledge networks self-organize into groups, which can also lead to links and communication clusters that become "echo chambers" for common views in a way that could hamper innovation.

Intellectual echo chambers run the risk of becoming over-specialized in a problem set. This can increase costs, limit value, and reduce the opportunity for success. Studies have shown that one way to reduce the chances for this kind of inertia is to ensure that clusters include diverse members across genders, age, and ranges of experiences. Knowledge networks are good at exchanging information and in introducing unlikely connections through "the strength of weak ties," as Mark Granovetter (1973) has termed it. We know that self-organization can be highly efficient for knowledge-creating communities, but it is also true that social connections may impose commitments that result in these groups outlasting their effectiveness. Diversity and links to problem spaces can help to reduce the liklihood of this problem by providing the demand for ideas and solutions that are not part of the connected, self-organizing network.

Network analysis and information technologies provide us with the tools to potentially enhance the effectiveness of knowledge-creating activities. The inertia within groups may emerge as a form of entropy, in which teams outlast their effectiveness. Anomalies arise and conditions of research teams change. These anomalies and conditions can be viewed within the network and possibly resolved earlier by using the knowledge we have of network dynamics. For example, we can view the network structure to see where high entropy conditions are arising. In these cases, we may find that a small investment of resources may recombine the teams into more fruitful configurations that produce better outcomes.

Further, complex systems often have fractal features—this is where the local, or subset, of a network reproduces the dynamics of the system as a whole. The question of whether the global network of science has fractal properties needs to be studied further, but it is possible to begin to test it by making local connections to the global system to see if dynamics can be tapped to enrich local capacities. If the local level is losing out to the global system, can the local nodes align themselves to the global in a way that helps transfer knowledge? This needs further exploration.

Along the lines of exploration, we know that the network structure changes the roles of individual nodes in the network. Collaborators can be mediators—playing the role of connecting one node or cluster to another one, or taking advantage of the weak links to make new connections. Second, other players are collective problem-solvers or people who join with others to ideate and link ideas to problem spaces. Third, exploratory actors reach far afield of their surrounding network to find and connect to utterly new and experimental ideas—often leading the way toward new sub-disciplines. A fourth actor is the resolver, the one who applies knowledge to a problem space. A fifth actor is the recipient who takes and uses a solution. Network structures mean that these roles can reconfigure frequently: today's mediator can be tomorrow's resolver, and so on.

Data mining algorithms can help here. Predictive data mining algorithms can track connections. Probabilities formulated as a result of network dynamics can track the knowledge system, helping recipients to query the system for connections. The recipients can test and ensure that solutions and ideas work, and quickly provide feedback to the knowledge creators. Recipients can trigger a "hot spot" where knowledge is needed. An agglomeration of knowledge creators can trigger a "hot spot" where investment in translation and dissemination may be fruitful. These capabilities are well within our reach, even now.

Policy changes can be and will be modeled through simulations of what we know about the global network of science, to identify the kinds of interventions and incentives that reduce entropy and that link more quickly to knowledge. Networks can be reconfigured by shifting the incentives, and this needs more research and understanding to ensure that the knowledge utility provides the goods that people need. Using the THICK methodology, it is possible to identify gaps in the system that may hinder knowledge exchange and diffusion.

Thus, new insights into common problem spaces can be derived from meta-knowledge that relates the global network of science to local demands and problems with a feedback loop—just as agricultural extension services did—but the knowledge utility will have vastly more complex and need more robust interactions. The ability to provide feedback is already in our hands—because so much of the world is connected on mobile devices. (Just as with any new technology, the first adopters are often criminals and marketers, but the use of feedback through (sometimes anonymous) mobile devices is in its infancy.) This system can be used to build the knowledge utility and provide fast feedback, but the infrastructure (software) for it will likely be provided as a public good, along the lines of Ostrom's global public goods vision.

Many experiments are needed to find out how best to impact the local economy with the knowledge utility infrastructure. The technology provides us the possibility of providing ease of access, workforce links, and ready-to-use training regardless of where the source and user of the information reside at a moment in time. The institutions and infrastructure to disseminate scientific knowledge will allow new economic models to emerge that may turn out to be radically local—linked locally, connected globally—as has been envisioned by Joseph Stiglitz. Radically local solutions have the additional advantage of contributing to social capital, enabling a virtuous cycle for a smart, caring community.

The coming of additive manufacturing, as one example, joined with the revolution in new materials, has been super-charged by the knowledge utility of problem and solution spaces—centers are offering the possibility of invigorating local communities toward sustainable and individually focused economic development. The potential for enhanced justice offered by the recombination of these factors at the local level is a vision of transformation of technology to the human scale. It is within our reach to create a knowledge utility infrastructure, but not without vision and policy to make it happen. As this vision unfolds, it has the counterintuitive possibility of helping us to disengage from machines and reconnect people to one another. Machines that are vastly "smarter" and connected, that help to provide the knowledge utility in real time, remove the need for human intervention between them. This disentangles people from machines and lets machines move into the background to provide the infrastructure that enhances human activity rather than mediate it, as they do now.

Indeed, science is difficult to govern, even at the national level, for all the reasons we have discussed: elitism, specialization, asynchronous results, loose coupling to outcomes, and esoteric vocabulary. Attempting to "plan" for scientific breakthroughs is a non-starter: no one really knows what is coming next in science. Attempts to create accountability regimes or return-on-investment reports for the sciences have been tricky, and the results, unsatisfying. Agencies of government that fund science will often focus research requests around broad fields of science, like "chemistry," rather than attempt to examine the dynamics of the knowledge-creating system (the "black box") to better understand it. Agencies may issue requests for focused problem-solving solutions (such as the improvement of materials for solar panels) and fund a scatter-shot of approaches. Others fund the cream-of-the-crop top scientists in hopes that leading thinkers will be productive. These have been rudimentary efforts that can be greatly improved by understanding the network of scientific communications and the social dynamics of science.

Robust scientific research appears able to flourish under many types of governing authorities. Indeed, it does not appear to require democratic institutions in order to flourish. The fact that science and democratic institutions often co-exist may be more closely linked to economic growth than to democracy. As Amartya Sen (1999) has pointed out, democratic governments have been highly effective at helping enhance growth. Although they do not seem to be a core requirement, since some non-democratic states have successfully funded science. They also may be better at disseminating the results of science because of an environment of openness, but this advantage is likely reaching its natural end. Research can flourish under many types of governing structures, and those that can recombine and feedback the problems and solutions may find they have the advantage. The creation of a legitimate public order and some type of knowledge utility to convey knowledge does appear to be an essential condition.

Theory tells us that good governance involves consensus, community, legitimacy, organization, effectiveness, and stability. These apply to science in different ways than they would to providing national defense or developing transportation policy. Political consensus around science is only that science is good and that it produces useful outcomes. The details around what is funded, how it is funded, and who should control and have access to the results are contentious and variable issues that have been debated by policymakers, analysts, and scholars and should be open to experiments. Various approaches have been crafted under different political systems to fund science. The policymakers who come to understand networks will have the advantage going forward. The following sections describe the governance dynamics at the three levels of research activity.

GOVERNANCE DYNAMICS AT THE INDIVIDUAL LEVEL

Researchers operating in the collaborative era are finding that the system of reward and recognition at the global level is far more complicated than it was in the past. Rewards and recognition were built as systems of feedback and control for national, disciplinary, or place-based science, while the collaborative era adds the global network to the mix. To complicate things even more, the higher the reputation of the researcher, the more likely it is that they are working at the global level. As a result of these layered systems, the individual researcher working within the global network of science responds to many conflicting signals. Tensions between competition and cooperation dominate many decisions. Collaboration can enhance research, making it more creative and offering a wider audience. But, evaluation processes still favor single-authored or first-author position on publications—acting as a disincentive for collaborations.

As the global network has emerged, many more researchers co-author on scientific journal articles than was the case in the past. This teaming trend appears to be productive, since more connections among more authors helps to spread the readership and increases the likelihood of citations. Having more authors on the author line means that citation counts are shared among a number of people (and perhaps, countries)—how to explain one's individual contribution? Universities need to work on improving metrics for those researchers who coauthor with others so that credit can be allocated fairly. Moreover, there is only one place for a first and one for a last author on the author list, which means that many people are relegated to middle positions. (Some journals are allowing "two first authors," but this is not widespread.) The same can be said about grant proposals where multiple researchers are listed, but often just a single person is the Principal Investigator. A proposal may represent a novel idea with an interdisciplinary approach, but that very approach and diverse team can become an obstacle if the review panel does not have the requisite expertise to assess the work for multiple disciplines.

The number of outlets for research results has also proliferated. ResearchGate, arXivs, and Twitter offer an opportunity to reach a broader audience. Many more researchers use Google Scholar as a library, a dissemination platform, and a source of feedback. Should researchers make preprints available? Sharing of research results allows researchers to build upon the work of others, verify and validate data, minimize duplication, and get feedback. Research shows that different fields have varying practices, with the social sciences and mathematics fields more likely to share preprints than the natural or medical sciences (Thursby et al. 2018). The split appears to be across what Thursby et al. (2018) call the "NCC" variables—norms, competition, and commercialization orientation of a field. Those fields with higher chances for competition or commercialization are less likely to offer preprints or data.

The irony is that the researcher who works alone or in isolation will find he or she is hobbled in an attempt to gain resources to conduct further research. The researcher who collaborates will compete with others for reputation and reward. This basic conundrum of the collaborative era raises the stakes for researchers. On top of this, the network structure of global science intensifies the role of reputation, since preferential attachment to others is the dynamic driving the growth of the network. Protecting and enhancing reputation requires actively using networking tools. The attention to this effort further intensifies the risks of navigating between deeply conventional work displaying mastery, and risky research advancing the frontiers of knowledge.

The recognition and rewards system are designed as instruments of control for a professional science system operating at the laboratory level and within disciplines, rather than for networked, cross-disciplinary dynamics. The laboratory/disciplinary system has been largely been subsumed into less place-based and more dynamic network. The global network is determining the direction of research. This creates confusion for individual researchers seeking reputation. The levers of control no longer fit the emerging system, but evaluative measures for the networked system do not yet exist.

The aspect of networking that translates into an action plan for individual researchers is accessing the global network. We know that most collaborations begin face-to-face. So one part of the action plan requires the individual researcher to participate in international conferences in order to meet with possible collaborators. Proposals for grant funds should include a request for funds, to participate and present at international conferences. This is one of the most effective ways that researchers become known to one another. The other is meeting at a research site, in cases where research is place-based, such as a rain forest, glaciers, oceans and lakes, and other geodic collaboration. Other begin on a university-based exchange.

While random connections can lead to fruitful partnerships, they are rare. Connecting with "friends-of-friends" is usually more productive in a network. Importantly, it can be helpful to know in advance of any conferencing event or research site visits, what resources, knowledge base, and complementary skills the other person might have. Moreover, it is most helpful to find someone who has a slightly better visibility in the network than oneself. Publishing with another researcher who has a higher visibility can help bring attention to one's work. Moreover, linking with a researcher with slightly higher visibility or reputation ensures that the relationship is reciprocal—that the power relationship is reasonably balanced—which can greatly enhance the collaborative effort. A large inequity in reputation between cooperative researchers can distort the relationship and sometimes leave the less visible researcher worse off.

No single indicator can provide insight to assess the productivity of the individual researcher. A collection of indicators is needed, and these should be customized to the type of work being conducted. This need for guidance was the motivation for the writing of the Leiden Manifesto for research metrics, which suggests that both quantitative and qualitative evidence are needed to evaluate performance. The Leiden Manifesto (Hicks et al. 2015) has developed a list of goals for evaluating research output:

1. Quantitative evaluation should support qualitative, expert assessment.
2. Measure performance against the research missions of the institution, group, or researcher.
3. Protect excellence in locally relevant research.
4. Keep data collection and analytical processes open, transparent, and simple.
5. Allow those evaluated to verify data and analysis.
6. Account for variation by field in publication and citation practices.
7. Base assessment of individual researchers on a qualitative judgment of their portfolio.
8. Avoid misplaced concreteness and false precision.
9. Recognize the systemic effects of assessment and indicators.
10. Scrutinize indicators regularly and update them.[2]

Researchers also need to be aware that international projects face higher levels of complexity resulting in higher transactions costs, such as the costs of coordination and communication. The higher complexity of these projects can include the (1) awkwardness of working across time zones, (2) the

[2] Hicks, et al., Bibliometrics: the Leiden Manifesto for research metrics. *Nature, 520,* 7548, 22 April 2015.

need to travel periodically over long distances to work together, (3) the loss of information due to suboptimal communication routines, and (4) clashes of management systems (e.g., Leung 2013; Jeong et al. 2012). Any one or combination of these obstacles may suppress otherwise creative or atypical knowledge pairing behavior, as international participants may withhold differences of opinion and defer to a lead author. In this sense, international collaboration may lean toward more hierarchical governance centralized around single or fewer leaders—and the researchers need to be on guard against this, since it will inhibit creative work. Different worldviews, nomenclatures, languages, and expectations can have the effect of slowing the integration of ideas, and may encumber the quality and validity of the results. Dealing with these obstacles explicitly at the start of a project can increase chances of high quality output. Diverse backgrounds can also bring creative differences, as well.

GOVERNANCE DYNAMICS AT THE INSTITUTIONAL LEVEL

The governance dynamics at the institutional level are complicated by the emergent nature of the global network. The most reputed researchers within research institutions are working at the global level. This brings attention and, with it, attracts resources to conduct further research. The allure of working at the international level attracts the highly productive researchers. They are "chosen" to work within the global system based upon the resources they bring and their reputation. Oftentimes, they have had a side-by-side connection to a foreign researcher that continues when one of the researchers returns home. The prominence and visibility of the collaboration brings with it a higher reputation for the institution in rankings.

The challenge for any institution operating within the global network, or that wishes to operate within it, is to create policies that strike a balance between global connection and local interaction. On one hand, researchers are generally good at identifying people with whom to collaborate. They should not be discouraged from doing so. This kind of connection should be supported. However, researchers who are spending time cooperating at the global level may not be available at the local level. This can, at times, result in impoverishing the local level. Researchers who are putting in the time to work at a distance may be less available to work locally with students, fellow researchers, or local or regional business.

Global connections should be complemented with local connections, too. The diffusion of knowledge that comes from connections to leading edge research can be critical for education and knowledge creation at the local level. The feedback from the local level can often enrich the global level as well. This can be accomplished with policies that provide avenues to support the globally connected researcher with local links. The range of options for this action is so broad that there is no point in detailing these actions here. Different institutions will have their own culturally defined methods of conducting outreach and connectivity. Actions can include providing additional doctoral students to work with a prominent researcher to ensure that know-how is diffused, supporting symposia, offering online blogging opportunities, funding local workshops, or creating online platforms for knowledge sharing.

A related consideration is one where the institution seeks to attract researchers to visit to conduct research onsite. This can be an expensive option if the host institution is offering funds to visitors. If visitors have their own funding, then the costs would be to ensuring that equipment, laboratory space, or access to a unique resource (such as a collection or a piece of equipment) is attractive to a visitor. Offering incentives to visitors such as awards or funds for research assistance can be bonus. Research shows that researchers who visit other institutions, especially across countries, have a higher-than-average citation count—so the benefits can be strong for the researcher and the host institution.

GOVERNANCE DYNAMICS AT THE NATIONAL LEVEL

The global network of science has emerged over and above the national-level systems. In many ways, the global level sets the agenda for scientific and technological research. The national systems, which set priorities related to missions, and which provide funds to research institutions and researchers based upon merit, have very little influence on the global level. Funds from national and regional entities fund research, but they do not always choose which projects involve international collaboration.

Megascience projects such as the International Space Station or the Large Hadron Collider are the most visible form of international collaboration, but they are only one small part of the funding of these activities. The budgets set aside to provide support to capital funding are a small part of the budgets of countries supporting these projects. Then, researchers submit proposals to conduct research using these facilities, such as putting an additive manufacturing machine on the ISS or conducting a specific experiment on a synchrotron. These projects often result in multi-authored publications with addresses from many countries, as visiting researchers likely put the name and address of their home institution. Very few researchers are permanently stationed at large international facilities.

Research that focuses on specific natural resources often becomes the subject of international collaborations. These projects can benefit from collaborations, but the funding for cross-national collaborative work can be a problem if each researcher is separately funded. Coordinating these activities, when funding cycles have different timing, can be an obstacle to collaboration. Finding ways to fund two or more researchers from different institutions or countries could help ensure more fruitful collaborations.

Governments have funded science and technology research with the hopes of building a knowledge-based economy, innovative capacity, and economic growth. National policies have a nearly university confidence in the developmental benefits of education, and STEM education in particular. It is common for nations to set goals of additional S&T spending—this is especially true for developing countries that appear to view this investment as a sign of advanced development. The World Bank, for example, shifted in the late 1990s from institution building to begin emphasizing the role of highly trained STEM workers to national development. Far beyond economic development, or even social goals such as health, science has become increasingly linked to broader goals such as sustainable and resilient development. These shifts have implications for governance that go beyond questions of what percentage of funding should go toward R&D, or how many trained workers are needed. Here we see the idea of science as defining national identities closely tied to current and future policy.

The global network challenges the ways these goods are distributed. As science becomes a goal for more and more countries, available knowledge is more broadly dispersed. This looks on the surface like a positive change. However, it can also mean the nations find it more difficult to evaluate the benefits or the return on investment of their funding. This difficulty can also threaten science funding since some legislators will ask why national funds go to build international activities that have little return to the public coffers. Ensuring ways to demonstrate local loops can help stem this possibility.

Thus, the main action that can address this need is twofold: one is to develop better measures of national benefit from science no matter where it takes place. Second is to invest more in scanning globally and diffusing knowledge locally. Some governments have done this to great effect: the Japanese agency JICST has scanned the globe for good science for many years. Government-funded global scanning and local diffusion are not common practices, and they should receive more attention. Smart people are more often scattered across the globe. Good ideas and strong science takes place in many more locations than was the case in the past. Governments should dedicate more effort to keep national researchers aware and abreast of developments across the globe. Convening representatives from scientific societies and related businesses, identifying critical knowledge inputs needed, and identifying where in the world interesting and important work is taking place can be steps in this direction. This can be followed by a visit or virtual workshop to consult with others to seek opportunities for cooperation. Ensure that international projects include a plan for local dissemination and discussion.

REFERENCES

Bush, V. (1945). *Science the Endless Frontier*. Washington, DC: US Government Printing Office.

Granovetter, M. S. (1973). The Strength of Weak Ties. *American Journal of Sociology, 78*(6), 1360–1380. https://doi.org/10.1086/225469

Hicks, D., Wouters, P., Waltman, L., Rijcke, S. D., & Rafols, I. (2015). Bibliometrics: the Leiden Manifesto for Research Metrics. *Nature, 520*, 429–431.

Jeong, S., Choi, J. Y., & Kim, J. Y. (2012). On the Drivers of International Collaboration: The Impact of Informal Communication, Motivation, and Research Resources. *Science and Public Policy, 41*(4), 520–531.

Leung, R. C. (2013). Networks as Sponges: International Collaboration for Developing Nanomedicine in China. *Research Policy, 42*(1), 211–219.

Mokyr, J. (2002). *The Gifts of Athena: Historical Origins of the Knowledge Economy*. Princeton: Princeton University Press.

O'Toole, L. J. (1997). Practical and Agendas in Public Administration. *Public Administration Review, 57*(1), 45–52.

Ostrom, E. (1990). *Governing the Commons*. Cambridge, UK: Cambridge University Press.

Price, D. J. (1963). *Big Science, Little Science*. New York: Columbia University Press.

Sen, A. (1999). *Development as Freedom*. New York: First Anchor Books.

Thursby, J., Haessler, C. Thursby, M., & Jiang, L. (2018, May 16). Prepublication Disclosure of Scientific Results: Norms, Competition, and Commercial Orientation. *Science Advances*. EAAR2133.

REFERENCES

Adams, J. (2010a). *Global science in perspective*. Conference paper.

Adams, J. (2010b). Science Heads East. *New Scientist, 205*(2742), 24–25.

Adams, J. (2013). Collaborations: The 4th Age of Research. *Nature, 497*, 557–560.

Ahmadpoor, M., & Jones, B. F. (2017). The Dual Frontier: Patented Inventions and Prior Scientific Advance. *Science, 357*(6351), 583–587.

Albert, R., & Barabási, A. L. (2002). Statistical Mechanics of Complex Networks. *Reviews of Modern Physics, 74*(1), 47.

Allen, R. C. (1983). Collective Invention. *Journal of Economic Behavior & Organization, 4*(1), 1–24.

Almeida, P., & Kogut, B. (1999). Localization of Knowledge and the Mobility of Engineers in Regional Networks. *Management Science, 45*(7), 905–917. https://doi.org/10.1287/mnsc.45.7.905

Amsden, A. H. (1992). *Asia's Next Giant: South Korea and Late Industrialization*. New York: Oxford University Press on Demand.

Archambault, E. (2010). 30 Years in Science: Secular Movements in Knowledge Creation. Science-Metrix. www.science-metrix.com/30years-Paper.pdf. Accessed July 2018.

Archambault, É., Beauchesne, O. H., Côté, G., & Roberge, G. (1988). Scale-Adjusted Metrics of Scientific Collaboration. Science-Metrix.com

Barabási, A., & Albert, R. (1999). Emergence of Scaling in Random Networks. *arXiv*:cond-mat/9910332v1, 1–11.

Barnard, H., & Pendock, C. (2013). To share or not to share: The role of affect in knowledge sharing by individuals in a diaspora. *Journal of International Management, 1*(12), 47–65.

© The Author(s) 2018

C. S. Wagner, *The Collaborative Era in Science*, Palgrave Advances in the Economics of Innovation and Technology, https://doi.org/10.1007/978-3-319-94986-4

Barnard, H., Cowan, R., & Muller, M. (2010). Global Excellence at the Expense of Local Relevance, or a Bridge Between Two Worlds? Research in Science and Technology in the Developing World. MERIT Working papers, No. 51. Maastricht: United Nations University.

Barnard, H., Cowan, R., & Müller, M. (2012). Global Excellence at the Expense of Local Diffusion, or a Bridge Between Two Worlds? Research in Science and Technology in the Developing World. *Research Policy, 41*(4), 756–769.

Barzel, B., & Barabási, A.-L. (2013, September). Universality in Network Dynamics. *Nature Physics, 9.* https://doi.org/10.1038/nphys2741

Beaver, D. deB, & Rosen, R. (1978). Studies in Scientific Collaboration – Part I. The Professional Origins of Scientific Co-authorship. *Scientometrics, 1*(1), 65–84. https://doi.org/10.1007/BF02016840

Benkler, Y. (2004). Sharing Nicely: On Shareable Goods and the Emergence of Sharing as a Modality of Economic Production. *Yale Law Journal, 114,* 273.

Bernal, J. D. (1939). The Social Function of Science. *The Social Function of Science.*

Blumenthal, M. S., Inouye, A. S., & Mitchell, W. J. (Eds.). (2003). *Beyond Productivity: Information, Technology, Innovation, and Creativity.* Washington, DC: National Academies Press.

BOAI. (2001). see: http://www.budapestopenaccessinitiative.org/

Borgman, C. L. (2010). *Scholarship in the Digital Age: Information, Infrastructure, and the Internet.* London: MIT Press.

Borgman, C. L., Wallis, J. C., & Enyedy, N. (2007). Little Science Confronts the Data Deluge: Habitat Ecology, Embedded Sensor Networks, and Digital Libraries. *International Journal on Digital Libraries, 7*(1–2), 17–30.

Boschma, R., & Frenken, K. (2011). The Emerging Empirics of Evolutionary Economic Geography. *Journal of Economic Geography, 11*(2), 295–307. https://doi.org/10.1093/jeg/lbq053

Bowker, G. C. (2008). *Memory Practices in the Sciences.* Cambridge, MA: MIT Press.

Boyack, K. W., Klavans, R., & Börner, K. (2005). Mapping the Backbone of Science. *Scientometrics, 64*(3), 351–374. https://doi.org/10.1007/s11192-005-0255-6

Bradford, S. C. (1985). Sources of Information on Specific Subjects. *Journal of Information Science, 10*(4), 173–180.

Braun, T., Glanzel, W., & Schubert, A. (2005). A Hirsch-style index for journals. *The Scientist, 19*(22). Letter, accessed Sept 2018, https://www.the-scientist.com/letter/a-hirsch-type-index-for-journals-48137.

Braun, T., Glänzel, W., & Schubert, A. (2006). A Hirsch-type index for journals. *Scientometrics, 69*(1), 169–173.

Bush, V. (1945). *Science the Endless Frontier.* Washington, DC: US Government Printing Office.

Cartwright, N., & Hardie, J. (2012). *Evidence Based Policy: A Practical Guide to Doing It Better.* Oxford: Oxford University Press.

Castells, M. (2011). *The Power of Identity: The Information Age: Economy, Society, and Culture* (Vol. 2). New York: Wiley.

Cetina, K. K. (2009). *Epistemic Cultures: How the Sciences Make Knowledge*. Cambridge: Harvard University Press.

Coe, N. M., Hess, M., Yeung, H. W., Dicken, P., & Henderson, J. (2004). "Globalizing" Regional Development: A Global Production Networks Perspective. *Development, 29*(4), 468–484. https://doi.org/10.1111/j.0020-2754.2004.00142.x

Cole, S., & Cole, J. R. (1967). Scientific Output and Recognition: A Study in the Operation of the Reward System in Science. *American Sociological Review, 32*(3), 377–390. https://doi.org/10.2307/2091085

Crane, D. (1972). *Invisible Colleges: Diffusion of Knowledge in Scientific Communities*. Chicago: University of Chicago Press.

Dahlman, C., & Aubert, J.-E. (2001). *China and the Knowledge Economy: Seizing the 21st Century*. Washington, DC: World Bank.

Darwin, C. (2004). *On the Origin of Species, 1859*. Routledge.

David, P. A. (2004). Understanding the Emergence of 'Open Science' Institutions: Functionalist Economics in Historical Context. *Industrial and Corporate Change, 13*(4), 571–589.

de Solla Price, D. J. (1963). *Little Science, Big Science*. New York: Columbia University Press.

Dodgson, M., Gann, D. M., & Salter, A. (2008). *The Management of Technological Innovation: Strategy and Practice* (Revised). Oxford: Oxford University Press.

Dolfsma, W., & Leydesdorff, L. (2009). Lock-In and Break-Out from Technological Trajectories: Modeling and Policy Implications. *Technological Forecasting and Social Change, 76*(7), 932–941.

Doudna, J. A., & Sternberg, S. H. (2017). *A Crack in Creation: Gene Editing and the Unthinkable Power to Control Evolution*. New York: Houghton Mifflin Harcourt Publishing Company.

Dunbar, R. I. (1992). Neocortex Size as a Constraint on Group Size in Primates. *Journal of Human Evolution, 22*(6), 469–493.

Florida, R. (2002). *The Rise of the Creative Class*. New York: Basic Books.

Florida, R. (2005, October). The World Is Spiky. *The Atlantic Monthly*.

Freeman, C. (1991). Technology, Progress and the Quality of Life. *Science and Public Policy, 18*(6), 407–418.

Freeman, C. (2002). Continental, National and Sub-national Innovation Systems – Complementarity and Economic Growth. *Research Policy, 31*, 191–211.

Freeman, C., & Hagedoorn, J. (1992). *Globalisation of Technology*. MERIT.

Frenken, K., Hoekman, J., Kok, S., Ponds, R., van Oort, F., & van Vliet, J. (2009). Death of Distance in Science? A Gravity Approach to Research Collaboration. In *Innovation Networks* (pp. 43–57). Berlin/Heidelberg: Springer.

Fukuyama, F. (1999). *The Great Disruption: Human Nature and the Reconstitution of Social Order*. New York: Free Press.

Garfield, E. (1971). *Essays of an Information Scientist* (Vol. 1, pp. 222–223), 1962–1973; Current Contents: #17.

Garfield, E. (1972). Citation Analysis as a Tool in Journal Evaluation. *Science, 178,* 471–479.

Garfield, E. (1979). Current contents: Its impact on scientific communication. *Interdisciplinary Science Reviews, 4*(4), 318–323.

Gass, W. H. (2006). *A Temple of Texts.* New York: Knopf.

Geim, A., & Novoselov, K. (2007). The Rise of Graphene. *Nature Materials, 6*(3), 183–191.

Gibbons, M., Limoges, C., Nowotny, H., Schwartzman, S., Scott, P., & Trow, M. (1994). *The New Production of Knowledge.* London: Sage.

Giddens, A. (1984). *The Construction of Society.* Cambridge: Polity.

Gilbert, G. N., Gilbert, N., & Mulkay, M. (1984). Opening Pandora's Box: A Sociological Analysis of Scientists' Discourse. CUP Archive.

Gokhberg, L., & Nekipelova, E. (2002). International Migration of Scientists and Engineers in Russia. In *International Mobility of the Highly Skilled* (pp. 177–187). Paris: Organisation for Economic Co-operation and Development.

Goncalves, B., Perra, N., & Vespignani, A. (2011). Validation of Dunbar's Number in Twitter Conversations, 8. Physics and Society; Other Condensed Matter; Human-Computer Interaction. https://doi.org/10.1371/journal.pone.0022656

Granovetter, M. S. (1973). The Strength of Weak Ties. *American Journal of Sociology, 78*(6), 1360–1380. https://doi.org/10.1086/225469

Guimera, R., Sales-Pardo, M., & Amaral, L. A. (2007). Classes of Complex Networks Defined by Role-to-Role Connectivity Profiles. *Nature Physics, 3*(1), 63.

Hagstrom, W. O. (1965). *The Scientific Community.* New York: Basic Books.

Harnad, S. (1997). The Paper House of Cards (and Why It's Taking so Long to Collapse). *Ariadne, 8,* 6–7.

Harrison, J. (1980). Hume's Theory of Justice. Oxford Scholarship Online: October 2011.

Harrison, C. E., & Johnson, A. (2009). *National Identity: The Role of Science & Technology.* Washington, DC: Osiris.

He, T. (2009). International Scientific Collaboration of China with the G7 Countries. *Scientometrics, 80*(3), 571–582. https://doi.org/10.1007/s11192-007-2043-y

Hess, C., & Ostrom, E. (2005). A Framework for Analyzing the Knowledge Commons (Draft). In C. Hess & E. Ostrom (Eds.), *Understanding Knowledge as a Commons: From Theory to Practice.* Cambridge, MA: MIT Press.

Hesse, M. (1963). A New Look at Scientific Explanation. The Review of Metaphysics, *17,* 98–108.

Hesse, M. (1974). *The Structure of Scientific Inference.* London: Macmillan.

Hicks, D., Wouters, P., Waltman, L., Rijcke, S. D., & Rafols, I. (2015). Bibliometrics: the Leiden Manifesto for Research Metrics. *Nature, 520,* 429–431.

Holland, J. H. (1995). *Hidden Order: How Adaptation Builds Complexity.* Reading: Perseus.

IMD. (2009). *IMD World Competitiveness Yearbook.* Lausanne: World Competitiveness Center.

Jacobs, J. (1969). *The Economy of Cities.* New York: Columbia University Press.

Jasanoff, S. (Ed.). (2004). *States of Knowledge: The Co-production of Science and the Social Order.* London: Routledge.

JCR. (2008). *Journal Citation Reports* (Science Edition). Clarivate Analytics.

Jeong, S., Choi, J. Y., & Kim, J. Y. (2012). On the Drivers of International Collaboration: The Impact of Informal Communication, Motivation, and Research Resources. *Science and Public Policy, 41*(4), 520–531.

Jinha, A. E. (2010). Article 50 Million: An Estimate of the Number of Scholarly Articles in Existence. *Learned Publishing, 23*(3), 258–263. https://doi.org/10.1087/20100308.

Jones, C. I., & Williams, J. C. (1997). Measuring the Social Return to R&D. Federal Reserve.gov

Juma, C. (2011). *The New Harvest: Agricultural Innovation in Africa.* Oxford: Oxford University Press.

Kauffman, S. (1996). *At Home in the Universe: The Search for the Laws of Self-organization and Complexity.* Cary: Oxford University Press.

Klein, J. T. (1996). *Crossing Boundaries: Knowledge, Disciplinarities, and Interdisciplinarities.* Charlottesville: University of Virginia Press.

Kontopoulos, K. M. (2006). *The Logics of Social Structure* (Vol. 6). Cambridge: Cambridge University Press.

Kuhn, T. (1962). *The Structure of Scientific Revolutions.* Chicago: The University of Chicago Press.

Kuhn, T. S. (2012). *The Structure of Scientific Revolutions.* Chicago: University of Chicago Press.

Kumar, A., & Asheulove, P. (2011). *Report on World Publications.* Conference Paper.

Larsen, P., & Von Ins, M. (2010). The Rate of Growth in Scientific Publication and the Decline in Coverage Provided by Science Citation Index. *Scientometrics, 84*(3), 575–603.

Latour, B. (1987). *Science in Action: How to Follow Scientists and Engineers Through Society.* Cambridge: Harvard University Press.

Lazer, D., Pentland, A. S., Adamic, L., Aral, S., Barabási, A. L., Brewer, D., et al. (2009). Life in the Network: The Coming Age of Computational Social Science. *Science (New York, NY), 323*(5915), 721.

Lemley, M. A. (2016). The Surprising Resilience of the Patent System. *Texas Law Review, 95*, 1.

Leung, R. C. (2013). Networks as Sponges: International Collaboration for Developing Nanomedicine in China. *Research Policy, 42*(1), 211–219.

Leydesdorff, L. (2001). *The Challenge of Scientometrics: The Development, Measurement, and Self-organization of Scientific Communications.* Boca Raton: Universal-Publishers.

Leydesdorff, L. (2015). The dynamics of journal-journal citation relations: Can "hot spots" in the sciences be mapped? *Proceedings of the Association for Information Science and Technology, 52*(1), 1–4.

Leydesdorff, L., & de Nooy, W. (2017). Can "Hot Spots" in the Sciences Be Mapped Using the Dynamics of Aggregated Journal–Journal Citation Relations? *Journal of the Association for Information Science and Technology, 68*(1), 197–213. https://arxiv.org/abs/1502.00229.

Leydesdorff, L., Wagner, C. S., & Bornmann, L. (2014). The European Union, China, and the United States in the Top-1% and Top-10% Layers of Most-frequently Cited Publications: Competition and collaborations. *Journal of Informetrics, 8*(3), 606–617.

Lichtenthaler, U., & Lichtenthaler, E. (2009). A Capability-Based Framework for Open Innovation: Complementing Absorptive Capacity. *Journal of Management Studies, 46*(8), 1315–1338. https://doi.org/10.1111/j.1467-6486.2009.00854.x

Lotka, A. J. (1926). The Frequency Distribution of Scientific Productivity. *Journal of Washington Academy of Sciences, 16*, 317–323.

Luhmann, N. (1982). *The World Society as a Social System.* Taylor & Francis.

Lundvall, B. A. (2009). Innovation as an interactive process: User-producer interaction to the national system of innovation. *African Journal of Science, Technology, Innovation and Development, 1*(2–3), 10–34.

Mabe, M. (2003). The Growth and Number of Journals. *Serials, 16*(2), 191–198.

MacKinnon, D., Cumbers, A., Pike, A., Birch, K., & McMaster, R. (2009). Evolution in Economic Geography: Institutions, Political Economy, and Adaptation. *Economic Geography, 85*(2), 129–150.

MacKinnon, D., Cumbers, A., & Chapman, K. (2010). Learning, Innovation and Regional Development: A Critical Appraisal of Recent Debates, *3*, 293–311. https://doi.org/10.1191/0309132502ph371ra

Maddox, B. (2002). *The Dark Lady of DNA.* London: HarperCollins Publishers.

Mancini, A. (2006). *International Patent Law Is Obsolete.* New York: Buenos Books America LLC.

Mann, M. E., Bradley, R. S., & Hughes, M. K. (1999). Northern Hemisphere Temperatures During the Past Millennium: Inferences, Uncertainties, and Limitations. *Geophysical Research Letters, 26*(6), 759–762.

Marshall, A. (2006). *Industry and Trade.* New York: Cosimo, Inc.

Merton, R. K. (1957). Priorities in Scientific Discovery: A Chapter in the Sociology of Science. *American Sociological Review, 22*(6), 635–659.

Merton, R. K. (1968). The Matthew Effect in Science. *Science, 159*(3810), 56–63.

Merton, R. K. (1973). *The Sociology of Science: Theoretical and Empirical Investigations.* Chicago: University of Chicago Press.

Moed, H. F., & Halevi, G. (2015). Multidimensional Assessment of Scholarly Research Impact. *Journal of the Association for Information Science and Technology, 66*(10), 1988–2002.

Mohrman, S., Galbraith, J. R., & Monge, P. (2004). *Network Attributes Impacting the Generation and Flow of Knowledge Within and from the Basic Science Community.* University of Southern California, Center for Effective Organizations.

Mokyr, J. (2002). *The Gifts of Athena: Historical Origins of the Knowledge Economy.* Princeton: Princeton University Press.

Monge, P. R., & Contractor, N. S. (2003). *Theories of Communication Networks. Computer* (Vol. 91). Retrieved from http://www.amazon.com/dp/0195160371

Morgan, K. J. (2007). The Learning Region: Institutions, Innovation and Regional Renewal, *31.* https://doi.org/10.1080/00343409750132289

Moravec, H. (1998). When Will Computer Hardware Match the Human Brain. *Journal of Evolution and Technology, 1*(1), 10.

Mulkay, M., Potter, J., & Yearley, S. (1983). Why an Analysis of Scientific Discourse is Needed. In K. Knorr & M. Mulkay (Eds.), *Science Observed: Perspectives on the Social Study of Science* (pp. 171–204). London: Sage.

Müller, K. H., & Riegler, A. (2014). Second-Order Science: A Vast and Largely Unexplored Science Frontier. *Constructivist Foundations, 10*(1), 7–15.

National Research Council. (2011). *Research-Doctorate Programs in the United States.* Washington, DC: National Academies Press.

National Science Board. (2011). *Science & Engineering Indicators.* Washington, DC: US Government Printing Office.

National Science Board. (2016). *Science & Engineering Indicators.* US Government Printing Office.

Nentwich, M., & König, R. (2012). *Science 2.0.* Frankfurt: Campus Verlag GmbH.

Nersessian, N. J. (1994). How Do Scientists Think? *Cognitive Models of Science.* Retrieved from http://www.cc.gatech.edu/aimosaic/faculty/nersessian/papers/how-do-scientists-think.pdf

Newman, M. E. (2001). The Structure of Scientific Collaboration Networks. *Proceedings of the National Academy of Sciences of the United States of America, 98*(2), 404–409. https://doi.org/10.1073/pnas.021544898

Nielsen, M. (2012). *Reinventing Discovery.* Princeton: Princeton University Press.

Nonaka, I., & Takeuchi, H. (1995). *The Knowledge-Creating Company.* New York: Oxford University Press.

Nonaka, I., & Toyama, R. (2003). The Knowledge-Creating Theory Revisited: Knowledge Creation as a Synthesizing Process. *Knowledge Management Research & Practice, 1*(1), 2–10.

Nonaka, I., Toyama, R., & Konno, N. (2000). SECI, Ba and Leadership: A Unified Model of Dynamic Knowledge Creation. *Long Range Planning, 33*(1), 5–34. https://doi.org/10.1016/S0024-6301(99)00115-6

Nonaka, I., Von Krogh, G., & Voelpel, S. (2006). Organizational Knowledge Creation Theory: Evolutionary Paths and Future Advances. *Organization Studies, 27*(8), 1179–1208.

O'Toole, L. J. (1997). Practical and Agendas in Public Administration. *Public Administration Review, 57*(1), 45–52.

OECD. (2014). *OECD Science, Technology and Industry Outlook 2014.* Paris: OECD Publishing. https://doi.org/10.1787/sti_outlook-2014-en.

Ogburn, W. F., & Thomas, D. (1922). Are Inventions Inevitable? A Note on Social Evolution. *The Academy of Political Science, 37*(1), 83–88.

Olson, M. (2009). *The Logic of Collective Action* (Vol. 124). Cambridge, MA: Harvard University Press.

Organization for Economic Cooperation and Development. (2018). *Main Science & Technology Indicators.* https://data.oecd.org. Accessed July 2018.

Ornstein, M. (1928). *Role of Scientific Societies in the Seventeenth Century.* Hamden/London: Archon Books (Reprint Edition, 1963).

Ostrom, E. (1990). *Governing the Commons.* Cambridge, UK: Cambridge University Press.

Padgett, J. F., & Powell, W. W. (2012). The Problem of Emergence. Chapter 1. In J. Padgett & W. Powell (Eds.), *The Emergence of Organizations and Markets* (pp. 1–29). Princeton: Princeton University Press.

Parsons, T. (1961). Some Considerations on the Theory of Social Change. *Rural Sociology, 26*(3), 219.

Polanyi, M. (1967). *The Tacit Dimension.* Chicago: The University of Chicago Press.

Powell, W. (2003). Neither Market nor Hierarchy. *The Sociology of Organizations: Classic, Contemporary, and Critical Readings, 315,* 104–117.

Powell, W. W., White, D. R., Owen-Smith, J., & Koput, K. W. (2005). Network Dynamics and Field Evolution: The Growth of Interorganizational Collaboration in the Life Sciences 1. *World, 110*(4), 1132–1205.

Price, D. D. S. (1961). *Science Since Babylon.* New Haven: Yale University Press.

Price, D. J. (1963). *Big Science, Little Science.* New York: Columbia University Press.

Price, D. J., & Beaver, D. (1966). Collaboration in an Invisible College. *American Psychologist* (July 1965).

Radosevic, S., & Yoruk, E. (2014). Are There Global Shifts in the World Science Base? Analysing the Catching Up and Falling Behind of World Regions. *Scientometrics.* https://doi.org/10.1007/s11192-014-1344-1

Raynaud, D. (2010). Why Do Diffusion Data Not Fit the Logistic Model? A Note on Network Discreteness, Heterogeneity and Anisotropy. In *From Sociology to Computing in Social Networks* (pp. 215–230). Vienna: Springer.

Rogers, E. M. (1995). *Diffusion of Innovations.* New York: Free Press.

Rogers, E. M. (2010). *Diffusion of Innovations.* New York: Simon and Schuster.

Rosenberg, N. (1982). *Inside the Black Box: Technology and Economics.* Cambridge: Cambridge University Press.

Science-Metrix. (2009). *30 Years in Science: Secular Movements in Knowledge Creation*. http://www.science-metrix.com/30years-Paper.pdf. Accessed Sept 2018.

Science-Metrix. (2018). Analytical Support for Bibliometrics Indicators: Open Access Availability of Scientific Publications. http://www.science-metrix.com/sites/default/files/science-metrix/publications/science-metrix_open_access_availability_scientific_publications_report.pdf. Accessed July 2018.

SCImago. (n.d.). *SJR — SCImago Journal & Country Rank* [Portal]. Accessed 2014, from http://www.scimagojr.com.

Scott, W. R. (1987). *Organizations*. Englewood Cliffs: Prentice Hall.

Sen, A. (1999). *Development as Freedom*. New York: First Anchor Books.

Shils, E. (1975). *Center and Periphery* (p. 3). Chicago: University of Chicago Press.

Simon, H. (1947). *Administrative Behavior*. New York: The Macmillan Company.

Simon, H. (1962). The Architecture of Complexity. *Proceedings of the American Philosophical Society, 106*(6). Retrieved from http://link.springer.com/chapter/10.1007/978-1-4899-0718-9_31

Small, H. (1973). Co-citation in the Scientific Literature: A New Measure of the Relationship Between Two Documents. *Journal of the American Society for Information Science, 24*(4), 265–269. https://doi.org/10.1002/asi.4630240406

Small, H. (1999). Visualizing Science by Citation Mapping. *JASIST, 50*(1973), 799–813.

Smith, A. (1937). *An Inquiry into the Nature and Causes of the Wealth of Nations*. Рипол Классик.

Stephan, P. E. (2012). *How Economics Shapes Science* (Vol. 1). Cambridge, MA: Harvard University Press.

Stichweh, R. (1996). *Science in the System of World Society*. Collected Papers on Niklas Luhman. *Social Science Information, 35*(2), 327–340.

Stokes, D. E. (2011). *Pasteur's Quadrant: Basic Science and Technological Innovation*. Washington, DC: Brookings Institution Press.

Taylor, M. Z. (2016). *The Politics of Innovation: Why Some Countries Are Better than Others at Science and Technology*. New York: Oxford University Press.

Thomas, L. (1978). *The Lives of a Cell: Notes of a Biology Watcher*. New York: Penguin.

Thursby, J., Haessler, C. Thursby, M., & Jiang, L. (2018, May 16). Prepublication Disclosure of Scientific Results: Norms, Competition, and Commercial Orientation. *Science Advances*. EAAR2133.

Toole, B. A. (1992). *Ada, the Enchantress of Numbers: A Selection of Letters from Lord Byron's Daughter and Her Description of the First Computer*. Sausalito: Strawberry Press.

UNESCO Science Report 2010. (2010). Geneva. Retrieved from http://www.unesco.org/new/en/natural-sciences/science-technology/prospective-studies/unesco-science-report/unesco-science-report-2010/

UNESCO Science Report Towards 2030. (2016). https://en.unesco.org/unesco_science_report?language=fr. Accessed June 2018.

Von Hippel, E. (1994). "Sticky Information" and the Locus of Problem Solving: Implications for Innovation. *Management Science, 40*(4), 429–439.

Wagner, C. S. (2009). *The New Invisible College: Science for Development.* Washington, DC: Brookings Press.

Wagner, C. S., & Jonkers, K. (2017). Open Countries Have Strong Science. *Nature News, 550*(7674), 32.

Wagner, C. S., & Leydesdorff, L. (2005). Network Structure, Self-Organization, and the Growth of International Collaboration in Science. *Research Policy, 34*(10), 1608–1618. https://doi.org/10.1016/j.respol.2005.08.002

Wagner, C. S., & Wong, S. K. (2011). Unseen Science? Representation of BRICs in Global Science. *Scientometrics, 90*(3), 1001–1013. https://doi.org/10.1007/s11192-011-0481-z

Wagner, C. S., Park, H. W., & Leydesdorff, L. (2015). The Continuing Growth of Global Cooperation Networks in Research: A Conundrum for National Governments. *PLoS One, 10*(7), e0131816.

Whitehead, A. N. (1925). *Science and the Modern World.* New York: New America Library, Macmillan.

Wagner, C. S., Park, H. W., & Leydesdorff, L. (2015). The Continuing Growth of Global Cooperation Networks in Research: A Conundrum for National Governments. *PLoS One, 10*(7), e0131816.

Whitley, R. (1984). *The Intellectual and Social Organization of the Sciences.* Oxford: Clarendon Press.

Wilson, E. O. (1999). *Consilience: The Unity of Knowledge* (Vol. 31). New York: Vintage Books.

Zhou, P., & Leydesdorff, L. (2006). The Emergence of China as a Leading Nation in Science. *Research Policy, 35*(1), 83–104. https://doi.org/10.1016/j.respol.2005.08.006

Zuckerman, H. (1967). Nobel Laureates in Science: Patterns of Productivity, Collaboration, and Authorship. *American Sociological Review, 32*, 391–403.

Zuckerman, H., & Merton, R. K. (1966). Patterns of Evaluation in Science: Institutionalisation, Structure and Functions of the Referee System. *Minerva, 9*, 66–100.

Index[1]

[1] Note: Page numbers followed by 'n' refer to notes.

© The Author(s) 2018   191
C. S. Wagner, *The Collaborative Era in Science*, Palgrave Advances
in the Economics of Innovation and Technology,
https://doi.org/10.1007/978-3-319-94986-4

Printed in the United States
By Bookmasters

Printed in the United States
By Bookmasters